Electrical Control Systems for Heating and Air Conditioning

Electrical Control Systems for Heating and Air Conditioning

by Clyde N. Herrick and Kieron Connolly

Published by
THE FAIRMONT PRESS, INC.
700 Indian Trail
Lilburn, GA 30047

Library of Congress Cataloging-in-Publication Data

Herrick, Clyde N.
Electrical control systems for heating and air conditioning by Clyde N. Herrick and Kieron Connolly.
p. cm.
Includes index.
ISBN 0-88173-277-X
1. Heating--Control. 2. Air conditioning--Control. 3. Heating--Electric equipment. 4. Air conditioning--Electric equipment.
I. Connolly, Kieron. II. Title.
TH7466.5.H45 1998 697--dc21 97-49003
CIP

Electrical control systems for heating and air conditioning by Clyde N. Herrick, Kieron Connolly.

Published by The Fairmont Press, Inc.
700 Indian Trail
Lilburn, GA 30247

Printed in the United States of America

10 9 8 7 6 5 4 3 2 1

ISBN 0-88173-277-X FP

ISBN 0-13-975194-7 PH

Distributed by Prentice Hall PTR
Prentice-Hall, Inc.
A Simon & Schuster Company
Upper Saddle River, NJ 07458

Prentice-Hall International (UK) Limited, London
Prentice-Hall of Australia Pty. Limited, Sydney
Prentice-Hall Canada Inc., Toronto
Prentice-Hall Hispanoamericana, S.A., Mexico
Prentice-Hall of India Private Limited, New Delhi
Prentice-Hall of Japan, Inc., Tokyo
Simon & Schuster Asia Pte. Ltd., Singapore
Editora Prentice-Hall do Brasil, Ltda., Rio de Janeiro

Table of Contents

Preface

Air conditioning, refrigeration and heating systems make use of complicated electrical devices and circuits to perform safety and control functions.

The technician must understand electrical principles, and be able to apply them to effectively troubleshoot and repair air conditioning systems. He/she must also understand the refrigeration cycle and the function of each subsystem within a refrigeration system.

This text has several purposes:

First, to provide the technician with an understanding of electricity and its inherent danger; and to present safe working practices.

Second, to present the theory of electricity and magnetism necessary to understand system components such as motors, contactors, relays, transformers and semiconductor devices.

Third, to assist the technician in understanding the interrelation of electrical and magnetic control devices in air conditioning, refrigeration, and heating systems.

Fourth, to present logical techniques for analyzing system problems and their repair.

Failure of a system creates inconvenience and economic hardship, whether it is a residential refrigerator-freezer, a commercial freezer in a supermarket, or an air conditioning system in a computer center. The customer expects an expedient return of system operation.

The final purpose of this text is to present the technician with the mental tools necessary to analyze a problem, come to a logical conclusion, and make the necessary repair.

Every effort has been made by the authors to utilize examples of systems that meet current standards.

We dedicate this text to our students, past, present and future. From our past and present students we have learned that the road to success is motivation, self esteem, and effort; although, not necessarily in that order.

Kieron M. Connolly
Clyde N. Herrick

San Jose City College

Chapter 1

Refrigeration And Air Conditioning

INTRODUCTION

The process of air conditioning includes air movement, air purification, changes in humidity, and changes of temperature. This book covers the concepts of the controls and electrical loads required to fulfill these factors. However, before studying the electrical concepts it is important that the technician review the concepts of air conditioning and refrigeration.

THE PROPERTIES OF AIR

An HVAC system moves "conditioned" air through a building. Conditioned air is air which is supplied in the correct amount, is clean, and has a temperature and humidity level desirable for both human comfort and effective operation of machinery and equipment. The quantity of air supplied to the area is dependent upon the speed and the size of the fan motor pushing the air, and the density and volume of the air. To effectively troubleshoot an air conditioning or refrigeration system the technician must understand the properties of air as well as electricity and refrigeration.

Air is a mixture of gasses and classified as either dry air or moist air. Dry air contains approximately 78% nitrogen, 20.9% oxygen, 1% argon, and 0.1% of other gasses. Moist air contains all of the above plus water vapor.

The *Specific Density* of air is defined as the weight of air for each one pound of space that it occupies. The *Specific Volume* of air is defined as the amount of space that one pound of air will occupy, and is measured in cubic feet per pound. The number of cubic feet per minute (CFM) of air that a fan or blower motor moves will have a bearing upon the amount of power and line current that the motor requires.

Another term often used in air conditioning is *Relative Humidity* which is the ratio of the amount of moisture present in air to the amount of water the air can hold before it becomes saturated. Relative humidity is expressed as a percentage. Air at 100% humidity contains all the moisture possible. Likewise, air at 50% humidity contains 50% of the moisture that it can contain.

The amount of moisture present in one pound of air is expressed in *grains*. *Specific Humidity* is the actual number of grains of moisture contained in one pound of air. In 1911 Dr. Willis Carrier presented his *Rational PSYCHROMETRIC formula* to the Society of Mechanical Engineers. His formula lead to the development of the "PSYCHROMETRIC chart" which is a graphical representation of the properties of air. The chart is very useful when determining the properties and conditions of air. However, it is of little use in the study of electricity and electrical controls. Therefore we will not elaborate on it here.

REFRIGERATION

Refrigeration is the removal of heat from a place where it is not wanted to a place where there is no objection to heat. When a substance is cooled, heat has been removed. Cold is the absence of heat. For example, it is necessary to remove heat from water to produce ice, to remove heat from air to condition it, and to remove heat from food products to preserve them. The electrical machinery used in the mechanical compression refrigeration cycle has one primary purpose—to remove and transfer heat. The technician must be familiar with the concepts of heat and heat measurements to understand the process of refrigeration.

THE PROPERTIES OF HEAT

Heat is a form of energy. Energy cannot be created nor destroyed. However, energy can be changed from one form to another. For example, a windmill generator changes wind energy to mechanical energy and then to electrical energy.

Heat can travel by one or a combination of the following methods:

- CONDUCTION is a method of heat transfer whereby heat is transferred directly from one object to another. For example, when an iron rod is placed in a flame, heat transfers from the flame to the rod by conduction.

- CONVECTION is the transfer of heat through the circulation of air. When heated gases or liquids expand and become lighter due to the reduction of their density, this causes them to rise and be replaced by heavier, cooler more dense gases or liquids. For example, boiling water converts to lighter steam and rises to produce currents called *convection currents.*

- RADIATION is a transfer of heat by energy waves called *radiation waves.* Radiation waves exist in an medium such as a vacuum. For example, the energy waves from the Klystron tube in a microwave oven pass throughout the interior. The substance in the oven absorbs the energy waves and converts them to heat.

Heat is a measurable quantity. The most common measurement is the Btu. *A Btu is the amount of heat required to raise or lower the temperature of one pound of water by one degree F of temperature.* The temperature scale used is the Fahrenheit scale.

Heat is not measured in degrees of temperature. Temperature is a sense of heat and does not tell us the amount of heat contained in a substance, but rather tells us how hot or cold we sense the substance. To clarify that statement: imagine that we have two containers of water, one containing 5 pounds and the other containing 1 pound. Suppose both containers are heated to 90°F. Both temperatures are the same; however, the 5 pounds of water has 5 times as much heat as the 1 pound of water. Both quantities of water sense the same temperature; however, the heat quantity is very different.

Example 1

How much heat must be added to 50 pounds of water to change its temperature from 70°F to 90°F?

Temperature rise = 70° – 90° = 20°

Btu change = 50 lb. × 20° = 1,000

Example 2

How much heat must be taken from 90 lb. of water to decrease its temperature from 100°F to 70°F?

Temperature decrease = 100° – 70° = 30°

Btu change = 90 lb. × 30° = 2,700

From the examples, we took away or added heat to cause a change of temperature. This type of change in heat is called *sensible heat.* Sensible heat is the form of heat which is added to or taken away from a substance to change its temperature *without* a change of state.

The three states of matter are *solids, liquids,* and *gasses.* A substance can have a change of state by adding or taking away heat from the substance. If we raise water to 212°F under normal atmospheric pressure, the water would boil and change to a gas. Likewise, if we reduce the temperature of the water to 32°F under the same conditions, the water would freeze and become a solid. The amount of heat added or taken from a substance that causes the substance to change states is called *latent heat.* Latent heat is the form of heat that when added or taken from a substance causes the substance to change states *without* a change in temperature.

The *boiling point* of a substance is the temperature at which the substance will change state from a liquid to a gas. The temperature is related to the pressure acting on the substance.

- If the pressure acting on a substance increases, the boiling point of the substance increases.

- If the pressure acting upon a substance increases, the temperature of the substance increases.

- If the pressure acting upon a substance is reduced, the boiling point of the substance is reduced.

- If the pressure acting upon a substance is reduced, the temperature of the substance is reduced.

Figure 1-1 is a graph of Temperature versus heat, Btu's per 1 lb. of water. It shows what happens to 1 lb. of water *under normal atmospheric pressure* when the heat applied is increased or decreased.

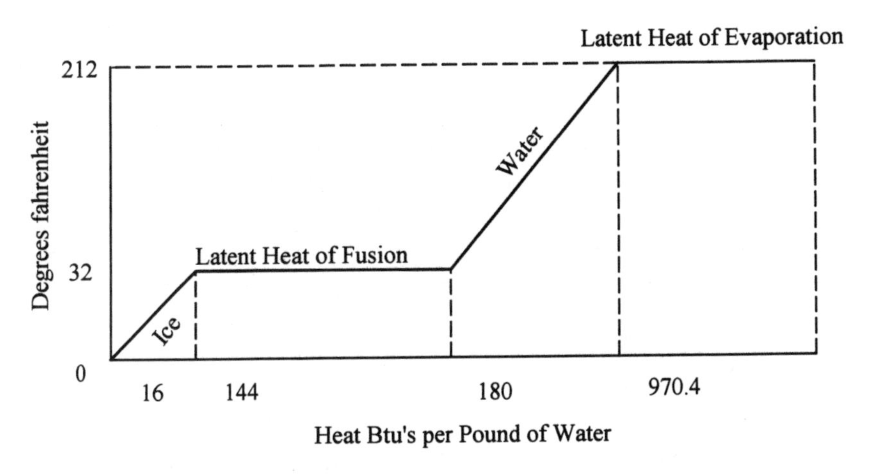

Figure 1-1. A graph of temperature versus heat.

The graph will be used as a reference in the study of an air conditioning unit. If a substance other than water were used in the example, the "specific heat" of the substance would have to be known. *Specific heat* is the ratio of heat required to change the temperature of a substance to the amount of heat required to change the temperature of an equal mass of water by the same number of degrees. The specific heat of a substance is found from a chart containing the specific heat of common substances. This is very helpful when designing a refrigeration system to decrease heat in a specific substance. Review the chart in Figure 1-1 and compare it to the facts you know about the boiling and freezing points of water.

THE STAGES OF REFRIGERATION

There are four stages of refrigeration that occur in the following sequence:

- **Compression-** A fluid such as a refrigerant is compressed by use of a compressor. Compressing the refrigerant raises its temperature and pressure. This now high-temperature, high-pressure refrigerant is discharged into a condenser.

- **Condensation-** The refrigerant vapor which has been discharged into the condenser is cooled to a temperature at which the gas changes into a liquid.

- **Expansion-** The liquid refrigerant enters a metering device which allows only a regulated amount of liquid through it. The pressure beyond the metering device is much lower; therefore, the liquid refrigerant passing through the device is exposed to low pressure.

- **Evaporation-** Beyond the metering device is the *evaporator,* which is under low pressure. This sudden pressure drop causes the liquid to boil off as a vapor, picking up heat from the evaporator in the process. The refrigerant vapor is drawn back into the compressor and the cycle repeats.

A mechanical compression refrigeration plant consists of a compressor, a condenser, a metering device, and an evaporator. Figure 1-2 is a block diagram of a refrigeration system. The diagram shows the changes that occur to the refrigerant contained in the system.

THE PRESSURE ENTHALPY DIAGRAM

At first glance the pressure-enthalpy diagram (sometimes called the Mollier diagram) appears complicated, but it is not hard to understand. It is a useful tool in analyzing refrigerant performance. The analysis is based on the refrigerant and is based on each pound of refrigerant contained in the system.

The chart is a graphical representation of the data contained in the

Figure 1-2. A block diagram of a refrigeration system.

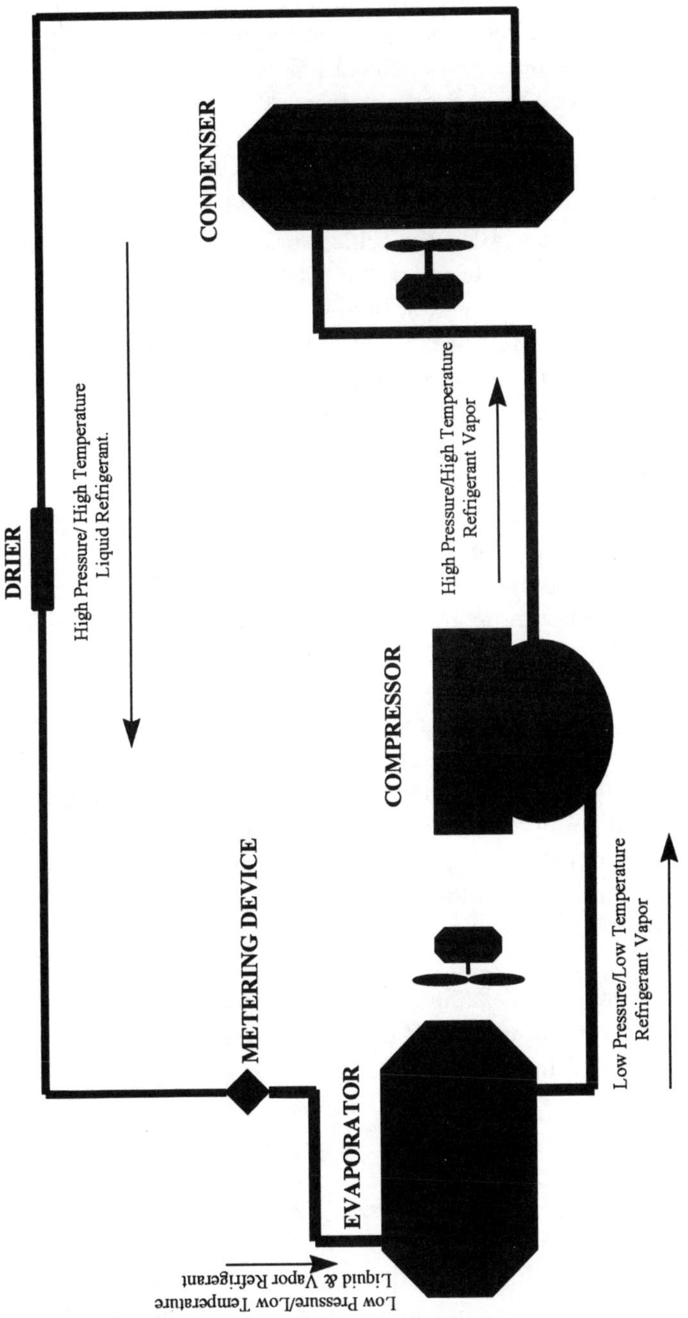

DRIER

High Pressure/ High Temperature
Liquid Refrigerant.

CONDENSER

High Pressure/High Temperature
Refrigerant Vapor

COMPRESSOR

METERING DEVICE

EVAPORATOR

Low Pressure/Low Temperature
Refrigerant Vapor

Low Pressure/Low Temperature
Liquid & Vapor Refrigerant

thermodynamic tables of individual refrigerants. The chart is divided into three zones. Each zone refers to the physical state of the refrigerant as it passes through the refrigeration cycle. Figure 1-3 shows the three zones. The zone on the left represents the subcooled liquid, the middle zone represents mixed vapor and the liquid state of the refrigerant and the right zone represents the refrigerant in the superheated state.

The boundary lines converge as pressure increases and come together at a point called the critical point. The critical point is the last point at which the refrigerant can exist. At temperatures higher than the critical point, the refrigerant can exist only in the vapor state.

The vertical line of the chart represents pressure in the unit of psia. Lines of constant pressure run horizontally across the chart. The pressure scale is logarithmic, which allows for large coverage on a small chart.

The horizontal scale represents enthalpy in units of Btu/lb. Lines of constant enthalpy are vertical. The enthalpy represents the energy content of each pound of refrigerant.

Lines of temperature run in a general vertical direction in the superheated vapor zone and the subcooled liquid zone. The temperature lines run in a horizontal direction in the mixed zone. Temperature lines are represented by the unit of degrees F.

The lines separating the zones show boundary conditions. The left boundary line is the saturated liquid line. Saturated liquid refrigerant exists on this line (that is refrigerant vapor with no liquid present). The right-hand boundary line is the line of saturated vapor (that is the refrigerant vapor with no liquid present). To the right of the saturated liquid line refrigerant is in both the liquid and the vapor state. To the left of the saturated vapor line the refrigerant is in both the liquid and the vapor state.

We will now examine the refrigerant flow through a system by drawing a refrigeration cycle on the pressure enthalpy diagram of the refrigerant used in the system. The cycle diagram consists of four lines which represent the effect of the major components of the mechanical system.

- **Condensation-** Condensation occurs at a constant pressure. A line drawn horizontally from right to left from the superheated vapor zone through the mixed zone and into the subcooled zone represents the process of condensation. This line is called the condensation zone. The length of this line in the superheated vapor zone indicates the amount of heat which must be removed before the vapor reaches saturation and the process of condensation can begin. An extension of this line into the supercooled liquid zone indicates removal of heat from the warm liquid. Cooling of the warm liquid starts near the outlet of the condenser and occurs all along the liquid line up to the metering device.

- **Evaporation-** Evaporation occurs at a constant pressure. A line drawn horizontally from the saturated liquid line through the mixed zone and into the saturated vapor zone is called the *evaporation line.* The length of this line in the mixed zone represents the increase in enthalpy at constant temperature and pressure—or the refrigerant pressure. An extension of this line into the superheated vapor zone indicates the amount of superheat added to the refrigerant on the low side of the system. Superheating of the refrigerant on the low side starts at the end of the evaporator and occurs along the suction line up to the suction inlet of the compressor.

- **Expansion-** Expansion occurs with no change of enthalpy. A line extending down from the condensation line to the evaporation line, crossing the saturation line at the temperature of the liquid refrigerant just before the metering device is called the *expansion line.* The lower end of this line tells us the percentage of refrigerant that has vaporized as the refrigerant leaves the metering device.

- **Compression-** Compression occurs at a constant enthalpy. A line extending upwards from the end of the evaporation line to the end of the condensation line is called the *compression line.* The compressed refrigerant increases in pressure, temperature, and enthalpy.

Figure 1-3 represents the history of a pound of refrigerant as it passes once through a system. The cycle represented illustrates a typical ideal saturated cycle that is based on the assumption that no vapor superheating occurs, no subcooling occurs, and no pressure losses occur except at the expansion device.

The actual refrigeration cycle differs from the ideal cycle in that su-

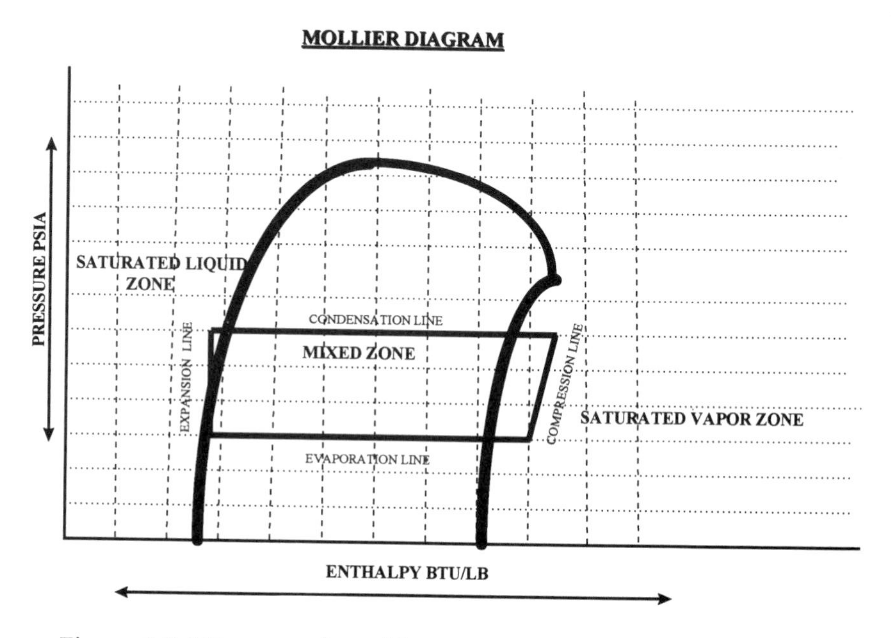

Figure 1-3. Diagram of a refrigerant as it passes around a system.

perheating of suction vapor normally occurs. Supercooling of the liquid refrigerant is normal. Pressure losses occur in every flow process as a necessary part of liquid flow. A small amount of heat flow occurs between the refrigeration lines and the surroundings (this can be significant on a long pipe run). Compression at a constant entropy can only be assumed for an ideal system. By adjusting the diagram to accommodate the above deviations, it is possible to analyze a running system and determine quantities such as :

— Condensation temperature.
— Evaporation temperature.
— Net refrigeration effect.
— Energy of compression.
— Heat rejection in condenser.
— Percent of vapor entering the evaporator.
— Coefficient of performance.

Example 3

The following measurements were taken from the table in Figure 1-4 for a system containing refrigerant R12.
— High side pressure = 180 psig.
— Low side pressure = 20 psig.

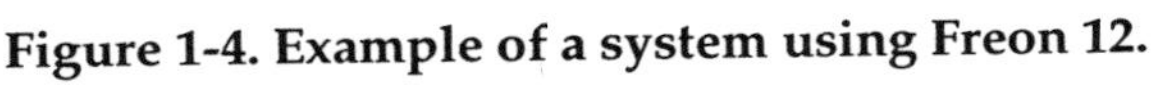

Figure 1-4. Example of a system using Freon 12.

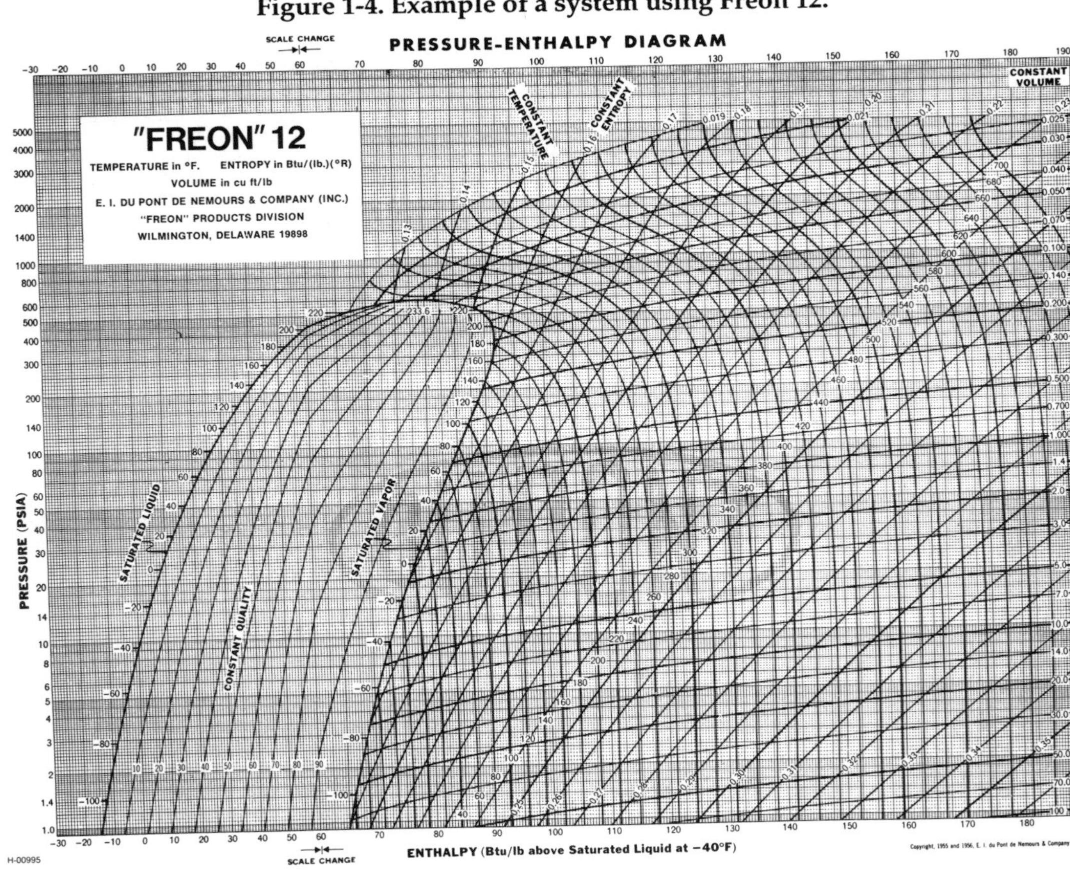

— Compressor discharge temperature = 190°F.
— Temperature of warm liquid line just before the metering device = 90°F.
— Temperature of the suction line just after the compressor = 46°F.

Find:
— Condensation temperature.
— Evaporation temperature.
— Net refrigeration effect.
— Energy of compression.
— High rejection of condensation.
— Percentage of vapor entering the evaporator.

Solution:
Convert the gauge reading to *psia*.
— High side pressure = 194.7 *psia*. Low side pressure = 34.7 *psia*.
— Draw the condensation line at a pressure of 194.7 *psia*.
— Draw the evaporation line at a pressure of 34.7 *psia*.
— Draw the expansion line vertically downward from the condensation line to cut the saturated liquid line at a temperature of 90°F.
— Mark the point at which the right end of the evaporation line intersects with the temperature of the suction line at 46°F.
— Mark the point where the right end of the condensation line intersects with the temperature of the high temperature refrigerant vapor at 190°F.
— Label the points A, B, C, D, and E as shown on the chart. This completes the diagram of the cycle.

The following can now be determined from the chart:
— *Condensation temperature* of the refrigerant is the temperature at which the condensation line intersects the saturated vapor line and saturated liquid line. = 128°F.
— *Evaporation temperature* of the refrigeration is the temperature at which the evaporation line intersects the saturated liquid line and saturated vapor line = 20°F.
— *Net refrigeration effect* is the difference in enthalpy between points E and A. = 58 Btu/lb.
— *Energy of compression* is the difference in enthalpy between points B

and A. = 15 Btu/lb.

— Heat rejection in the condenser and liquid lines is the difference in enthalpy between points B and A. = 80 Btu/lb.
— The percentage of refrigerant entering the evaporator is where the line of constant quality intersects point C.= 20%.

The Coefficient of Performance is calculated:

$$\text{Coefficient of Performance} = \frac{\text{Net Refrigeration Effect}}{\text{Energy of Compression}}$$

Note: This method of analysis illustrates the type of information available from the diagram and can be a useful tool. Occasionally, greater precision may be necessary than can be obtained from the diagram. In that case thermodynamic tables or computer software must be used.

In summary of the basic refrigeration cycle, we can say that liquid refrigerant will absorb heat from its surroundings and the refrigerant will change states from a liquid to a vapor. The vapor will be changed back into a liquid in the condenser. The temperature of condensation and evaporation of the refrigerant can be varied by changes of pressure.

REFRIGERANTS

A refrigerant is a substance that will pick up heat from another substance. On a warm day, drinking a glass of cold water will cool our body. Therefore, the water acts as a refrigerant. Ammonia, sulfur dioxide, and methyl chloride are gasses which were widely used as refrigerants in the past. Presently, refrigerants such as chlorofluorocarbon, hydrochlorofluorocarbons, and hydrofluorocarbons are commonly used. Due to stratospheric ozone depletion, refrigerants in the chlorofluorocarbon group have not been manufactured since 1996. The basic factors to consider when choosing a refrigerant are:

- Condensing pressures.
- Evaporation pressures.
- Latent heat of evaporation.
- Amount of heat per pound of refrigerant which the refrigerant is

capable of absorbing.

- Volume per pound of refrigerant.
- Flammability and toxicity of the refrigerant.
- Detectability.
- Corrosive action.
- Cost.
- Availability.

Refrigerant HCFC R22 is commonly used in most packaged air conditioning plants. In medium-temperature applications; such as reach-in coolers, walk-in coolers, and household refrigerators, refrigerant CFC R12 is commonly used. Since refrigerant R12 belongs in the CFC group which is no longer being produced, it is being replaced by HFC R134a which is more ozone friendly. Refrigerant R502, a mixture of R22 and R115 is commonly used in low-temperature applications such as blast freezers. However, R502 contains R115 which is a CFC refrigerant.

Never assume that a certain type of refrigerant is being used in a system. Simple methods of determining the type of refrigerant in a system are:

- Check the nameplate of the appliance.
- Examine the metering device (other than a capillary tube metering device).
- Use the pressure/temperature chart to compare the pressure of the refrigerant to the ambient temperature acting on it, and cross reference the type of refrigerant. Refer to Example 3.

Under no circumstances should refrigerants be mixed.

SUMMARY

- Before starting the study of electricity and electrical controls be sure to be familiar with the functions of COMPRESSORS, CONDENSERS, METERING DEVICES, AND EVAPORATORS.

- Understand the functions of OUTDOOR FAN MOTORS and CONDENSER FAN MOTORS, and understand their differences.

- Understand the difference between the open type, the semihermetic type, and the hermetic type compressors.

Chapter 2

Electrical Principles

INTRODUCTION

We use electricity dozens of times each day without the necessity of understanding it. We give little concern to where it comes from, how it is produced, or how its works. However, without electricity our daily routines would grind to a halt. Most of our conveniences would be lost: the electric stove, electric lights, hot water from the tap, radio, television, the automobile, trains, busses, planes, and traffic lights.

The layman need not be concerned about the theory of electricity. However, the technician must know how it works, for it is his or her job to keep the machines that depend on electricity in working order. The more one knows about electricity and electrical circuits the safer and better they can perform their job.

THE ATOM

Matter is any substance that has mass and occupies space. Matter can be in the form of a liquid, a solid or a gas. Elements are the basic form of matter. For example: oxygen, copper, and hydrogen are elements. On the other hand, water is a compound of hydrogen and oxygen.

The atom is the basic component of all matter in the universe, and all matter is made from a combination of atoms. Atoms are made up of neutrons, protons and electrons. Neutrons and protons of an equal number form the heavy center nucleus of the atom (Figure 2-1). The proton is said to have a *positive* charge.

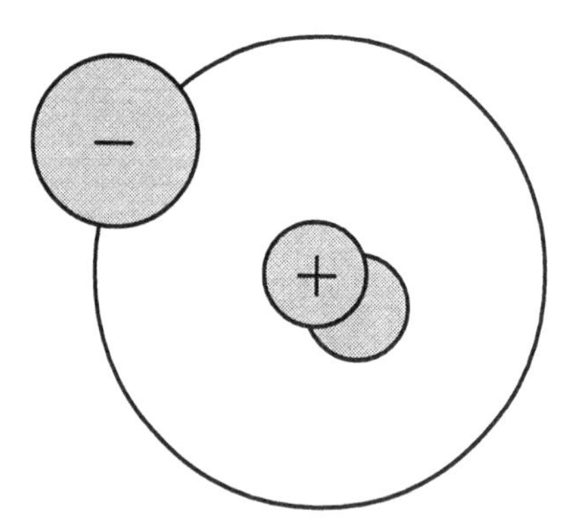

Figure 2-1. Principle parts of an atom.

Neutrons are a combination of a proton and an electron, and have a neutral charge because electrons have a negative charge. The negative charged electrons spin around the nucleus in rings or shells. The basic law of charges is that *like charges repel each other and unlike charges attract each other* as illustrated in Figure 2-2.

Figure 2-3 depicts an atom with ten electrons in two shells around the nucleus. The ten protons in the nucleus attract the spinning electrons

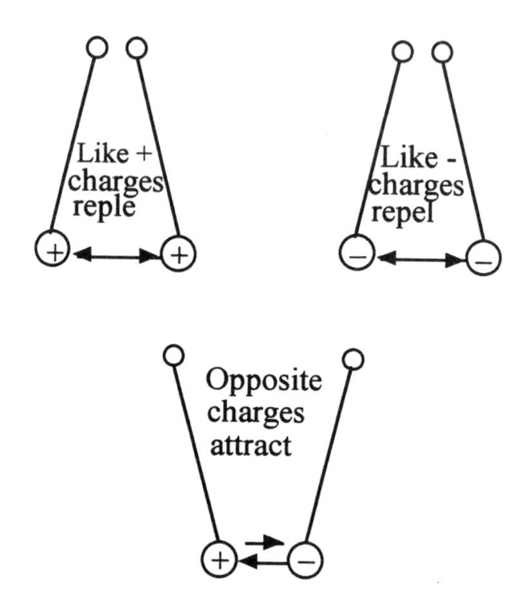

Figure 2-2. Attraction and repulsion of charges.

electrons and hold them to an orbit within the atom. The only parts of the atom that are of concern to us are the electrons in the outer shell. These electrons are called valence electrons. It is these electrons that become current flow in a circuit, and make materials such as wood, glass and rubber—insulators.

CONDUCTORS

Elements with one or two electrons in the valence band are unstable. With the addition of heat or voltage these electrons will be freed to become current flow. The freed electrons will move away from the negative terminal and toward the positive terminal of the source voltage. Current is the movement of electrons in a material. We say that electrical current which is the flow of electrons, flows from negative to positive. When an external voltage is applied to a material such as copper, gold, silver, and aluminum, the valence electrons are freed and current

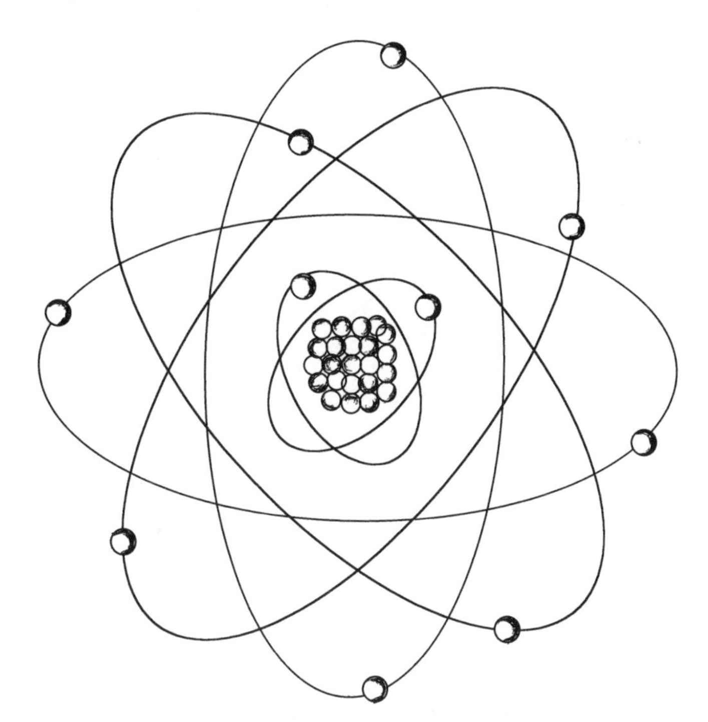

Figure 2-3. Example of an atom

flows easily. These electrons travel from atom to atom as shown in Figure 2-4.

VOLTAGE

Voltage is defined as an electromotive force (EMF), and is symbolized in formulas as the letters V or E. A voltage is a difference of potential of one point in respect to another that pushes and pulls electrons (current) through a conducting material.

STATIC ELECTRICITY

Static electricity is the result of a material losing or gaining electrons, through friction, as one material is rubbed against another. For example, on a dry day you may get a shock off a door handle after walking across a carpet. In this case, static electricity is produced by converting the friction of your shoes on the carpet into electrical energy. Static electricity is used in areas such as the coating of materials or removing particles from dirty air as in the operation of electronic air filters.

A simple demonstration of the laws of charges is to rub a glass or

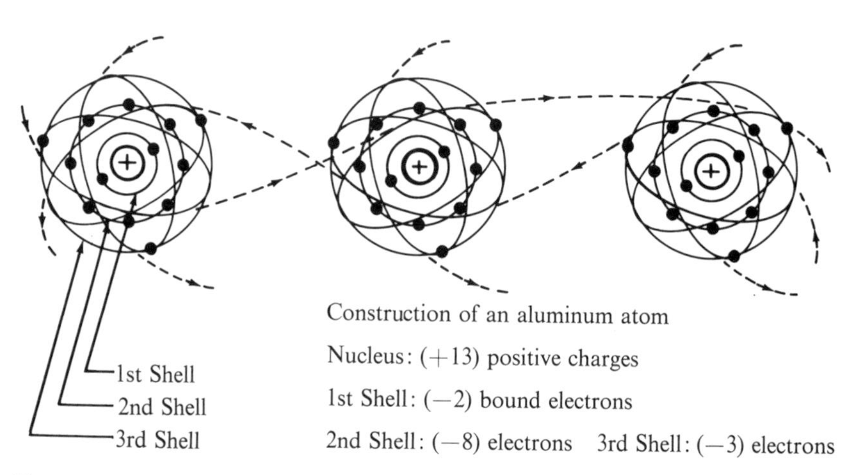

Figure 2-4. Current flow in a conductor is a movement of valence electrons.

plastic rod with a piece of fur (Figure 2-5). The demonstration can also be performed by passing a comb through your hair. As a comb is passed through the hair it picks up electrons (more electrons than protons) and is said to have a negative charge. The negatively charged comb will attract small pieces of paper, proving that unlike charges attract. The comb will hold the bits of paper until the charges become equal.

BATTERIES

When certain metals are placed in an electrolyte (acid) solution, one metal will give up electrons and the other will gain electrons. Some atoms lose electrons and some atoms gain electrons. This causes a difference of potential and forms a *battery.* Some wet-cell batteries, such as automobile batteries, can be recharged over and over. Others, called dry cells, Figure 2-6, cannot be recharged. All batteries convert chemical energy to electrical energy.

When two dissimilar metals are bonded together and heated at the tip, a difference of potential is produced and a current will flow. This device is called a *thermocouple.* Thermocouples convert heat energy to electrical energy and are used as current sources in some gas-fired furnaces. Figure 2-7(a) illustrates the thermocouple connection. Many configura-

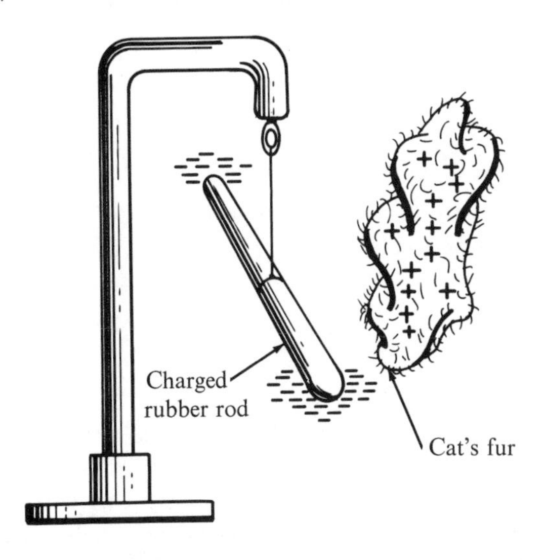

Figure 2-5. Static electricity is produced by friction.

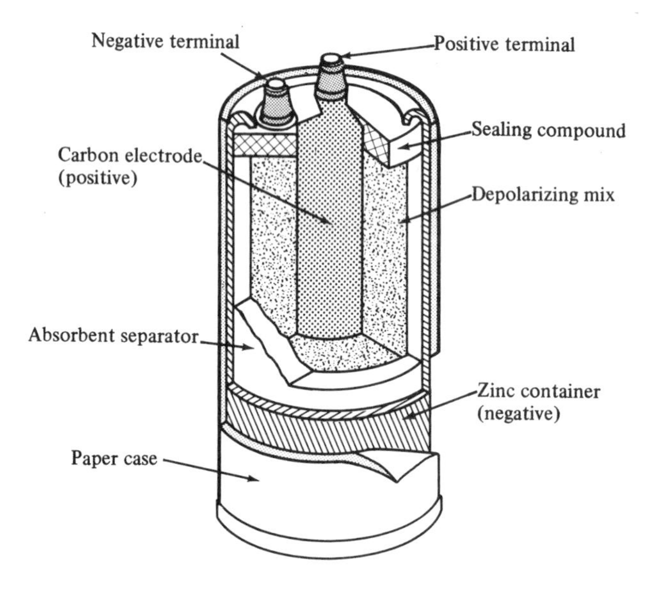

Figure 2-6. A cutaway view of a battery.

tions of thermostats are used in heating systems, Figure 2-7(b); however, all work on the same principle.

GENERATORS

When a conductor is moved through a magnetic field, a current is produced causing a voltage that can be used to perform work. The device for this conversion of mechanical energy to electrical energy is called a generator. Figure 2-7 depicts a basic generator with a coil of wire passing along a magnet.

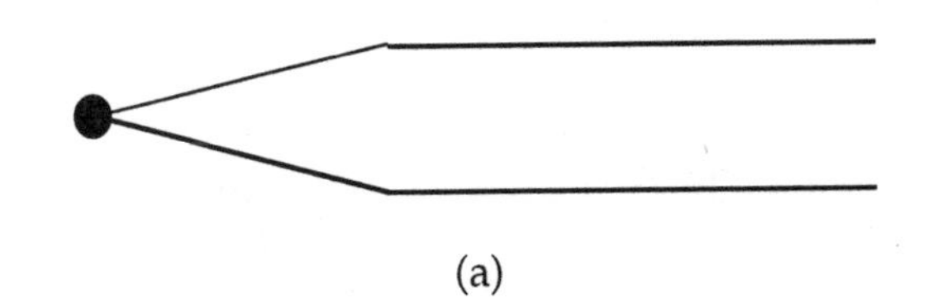

(a)

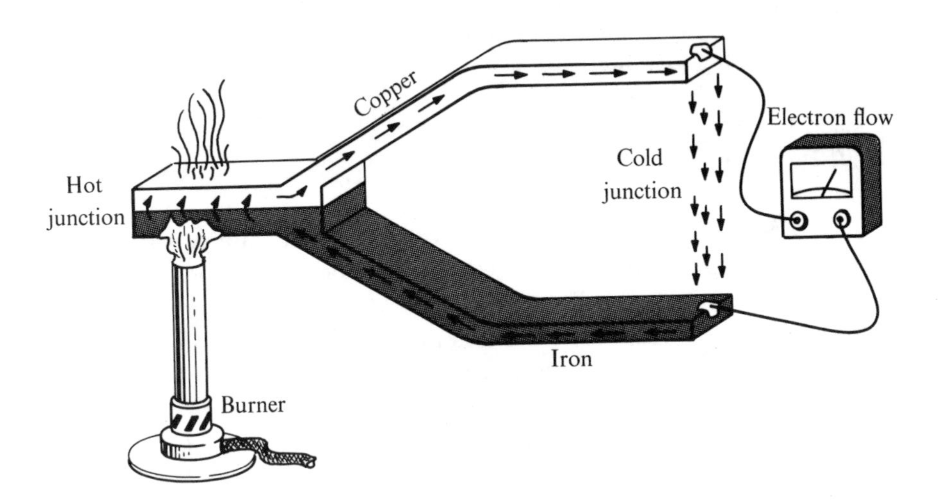

Figure 2-7. A thermocouple (a) circuit action, (b) symbol.

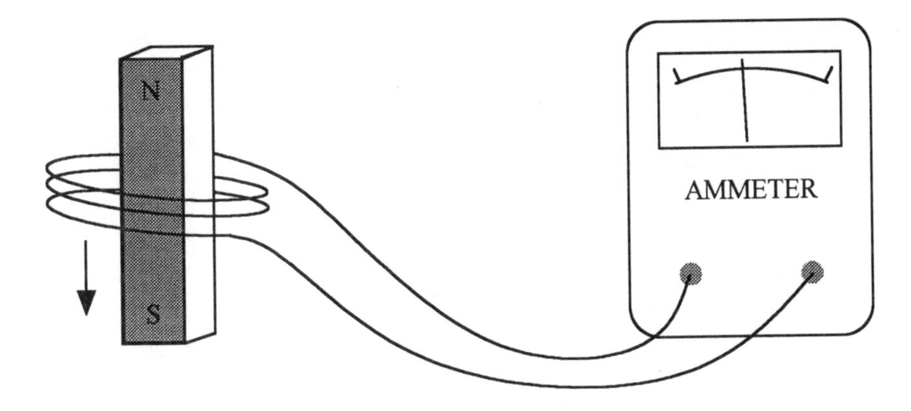

Figure 2-8. The movement of the coil over the magnet induces a current.

PHOTO CELL

Photo cells and solar cells produce a difference of potential when exposed to light or heat. Photo cells are useful in converting any light energy into electrical energy.

Solar cells are becoming efficient enough to be considered as a power source in some remote locations. Several solar-powered automobiles have made a trip across Australia; and we often see solar-powered telephones along our highways.

CURRENT FLOW

Current is the movement of electrons through a material as a result of a voltage. We define current as a certain number of electrons passing a point in one second. Specifically, 6.25 and 18 zeroes passing a point in one second is one ampere. Current is measured in *amperes* and is measured with an ammeter. The formula symbol used for current is the letter I. Current flows easily through materials that are good conductors. However, little current flows through materials that are insulators.

RESISTANCE

Resistance is the property of a material that limits current flow in a circuit. Resistance in a wire or other devices causes heat when a current is passed through them. All materials have resistance. For example, the filament of an incandescent light heats and glows because of resistance in the filament (Figure 2-9). Resistance is measured in *ohms* (Ω), and is referred to in formulas as R.

INSULATORS

Insulators are materials with few or no free electrons. Such materials are glass, ceramic, dry wood, rubber, ceramics, and most plastics. Insulators will break down by the application of an excessive voltage. Moisture, heat, and/or dirt cause most insulation breakdowns.

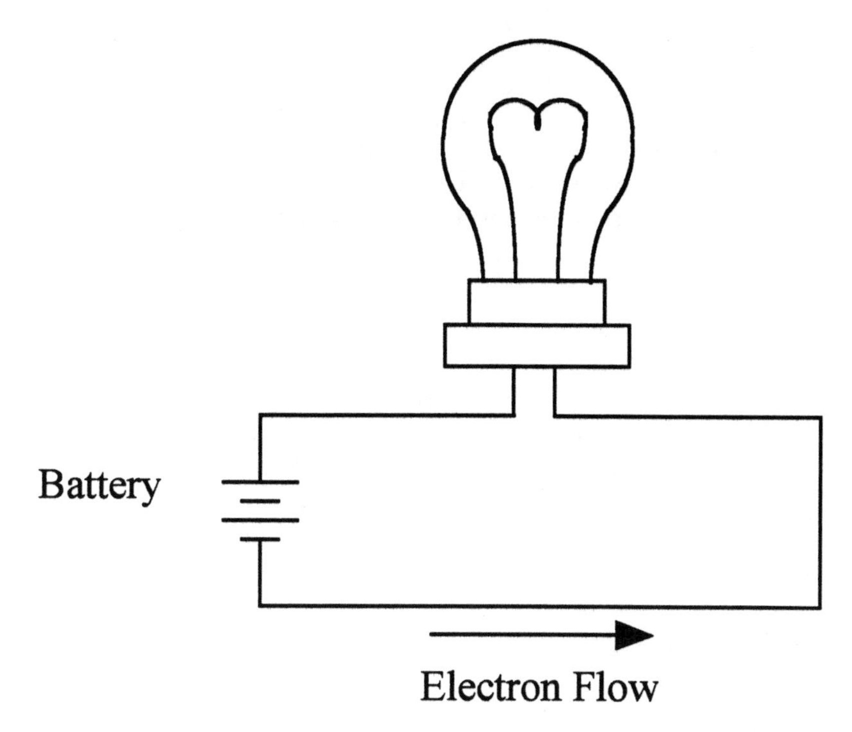

Figure 2-9. Resistance of the filament causes heat, which produces light, when a current is present.

SEMICONDUCTORS

Semiconductors are materials that are neither good conductors nor good insulators. The most common semiconductor materials are the elements germanium and silicon. Diodes and transistors are manufactured from semiconductor materials, usually silicon.

POWER

Power is the rate of doing work. In mechanical systems work is usually rated in *horsepower.* The rate of doing work in electrical circuits power is expressed in *watts.* For example, a 1000-watt heater is converting electrical energy into heat energy at the rate of 1000 watts per second.

ELECTRICAL INSTRUMENTS

Instruments are used to measure electrical quantities when testing components of environmental control systems. The levels of these measurements indicate fault-no-fault conditions of devices and equipment. The technician must become well-versed in the proper use of electrical instruments.

AC VOLTAGE

There are two types of electricity used in electrical circuits. The most common type for line or supply voltage is 110-volt, 208-volt, 220-volt, or 440-volt alternating voltages. These voltages are called alternating voltages (AC) because the voltage alternate above and below zero. Line voltage varies at a rate of 60 cycles per second, called Hertz (Hz). For example, a home service drop would be 220-V (60 Hz. Hertz is the measurement of the frequency of the wave. Figure 2-10 illustrated the voltage variations versus time for a 120 V 60 Hz AC voltage.

DC VOLTAGE

The second type of voltage and current is called direct current voltage (DC) because the current flows only one way in the circuit. DC voltage is used to operate circuits in control device modules that contain diodes, transistors and other semiconductor devices.

To operate the semiconductor devices in environmental control systems, the AC line voltage is converted to DC voltage.

THE AMMETER

The ammeter is used to measure the current in an electrical circuit. The ammeter depicted in Figure 2-11 is a *clamp-on ammeter* for measuring alternating current (AC). This ammeter is connected around *one wire* of a circuit to measure the AC current in that wire. It is sometimes necessary to amplify this reading. This is accomplished by looping the wire around the clamp and dividing the reading by the number of loops. This device cannot be used to measure DC current. DC clamp-on ammeters are available to measure high levels of DC current. Low levels of AC and

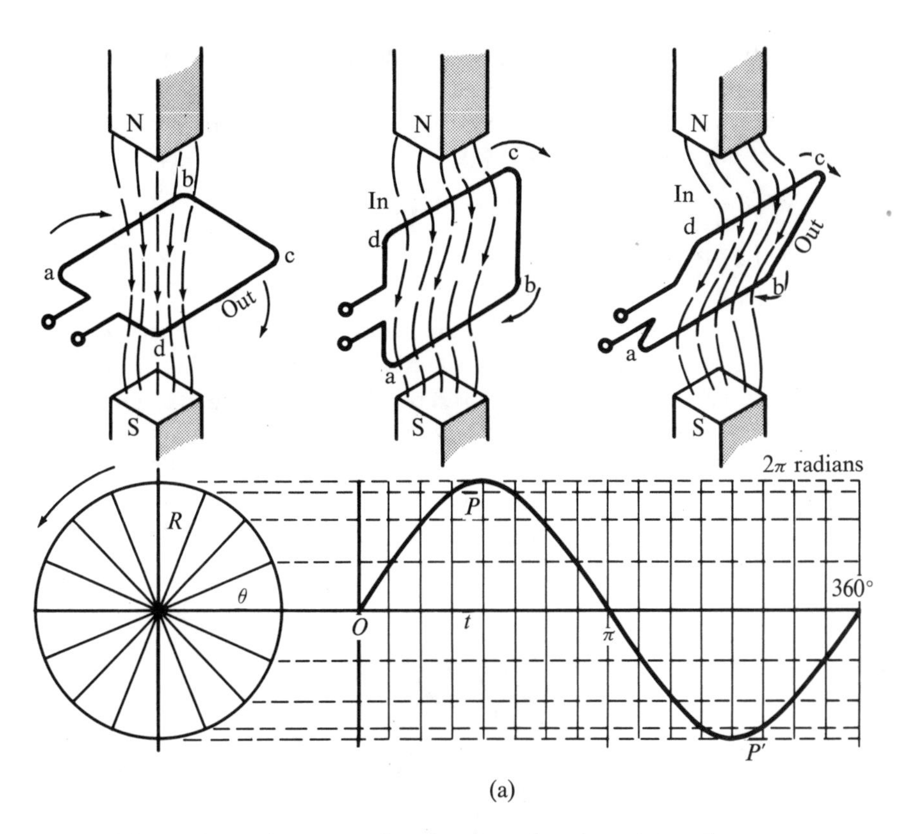

Figure 2-10. An example of one cycle of an AC voltage.

DC current are measured with an insertion type of ammeter. To use this type of instrument, the circuit is broken and the meter inserted in the break to complete the circuit. Air conditioning technicians seldom resort to this type of measurement, because this type of instrument can be easily destroyed and is difficult to install. The Digital Voltmeter (DVM) shown in Figure 2-12 can be used to measure low values of DC and AC currents. This instrument must always be connected in the circuit to measure AC or DC current as shown in Figure 2-13.

2.18 VOLTMETER

The voltmeter is always connected across or in parallel with the voltage to be measured. For example, the voltage at a 120-V AC receptacle would be tested across the two outlets. The voltage mode of AC or DC must be correctly selected for a correct voltage reading. A good rule is to always select a high range before connecting the meter and range down for a proper reading.

OHMMETER

Resistance is measured with an ohmmeter. Voltage must *always be removed* before connecting an ohmmeter. An ohmmeter has an internal voltage supply, and therefore, should be switched off when not in use. It is good practice to disconnect one end of the component being measured from the circuit to obtain a correct resistance reading. The DVM shown in Figure 2-11 contains an ohmmeter function. The ohmmeter can be used to test for breaks in a wire or an open or loose connector. This test is called a **continuity test.**

Figure 2-11. A clamp-on AC ammeter.

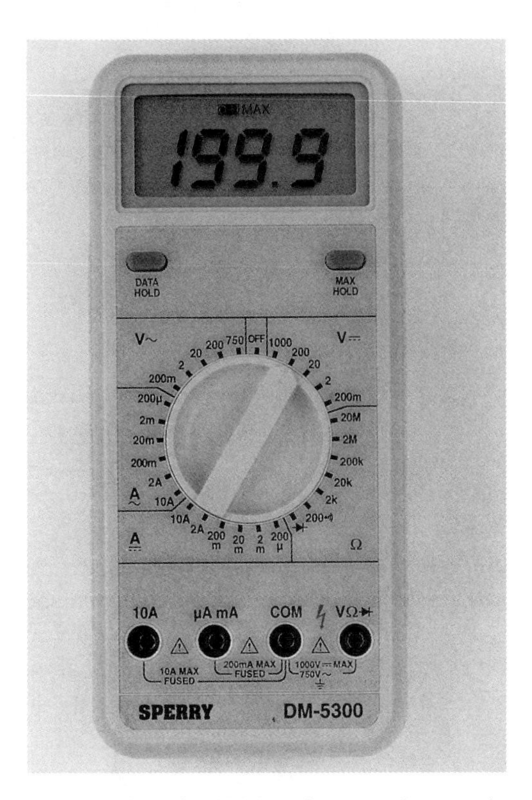

Figure 2-12. An example of a DVM that can be used as an insertion ammeter.

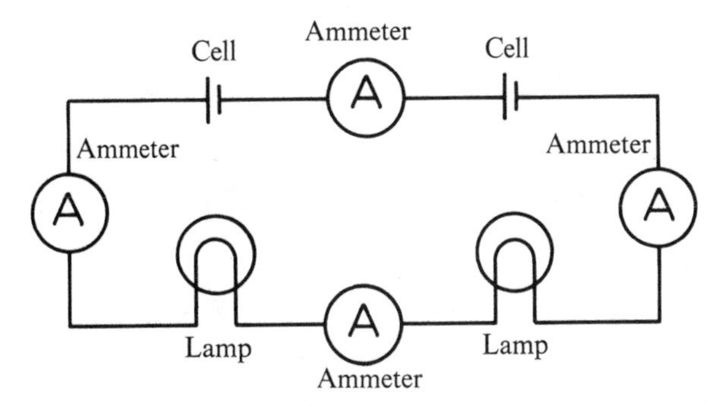

Figure 2-13. The connection of an insertion in-line ammeter to measure current.

WATTMETER

A Wattmeter can be used to measure the power consumption of a motor or other electrical load. The Wattmeter measures only the true power of the circuit. We will find later that it is sometimes necessary to measure the *apparent power.*

SAFETY

Every machine and every tool that uses electricity for its operation has the potential to kill. Electricity is our servant and is used to perform many tasks which make our life easier. However, it is always ready to cause injury or death to the user. *Always remember to have respect for electricity because electricity has no respect for you.*

You can be assured that sometime in your career as an air-conditioning technician, you will receive an electric shock. Understanding the nature of electricity and how it functions may save your life. A safety summation and test are presented in Addendum 1, for study and a self test. Here we will review only the most basic of electrical safety rules:

- If possible, never work on "live" equipment.
- Never intentionally take an electric shock. Even a low voltage can be dangerous.
- Never work on any equipment while standing on a damp or wet floor.
- Turn off main breakers or remove fuses and attach a warning tag.
- Always test for voltage on a circuit before touching—even a "dead" circuit.
- Never lean on the metal case of an equipment while testing a circuit.
- Never lean on another piece of equipment when testing a circuit.
- Always make your first touch of a circuit with the back of your fingers.
- Whenever possible test circuits with one hand and the other in your pocket.

- Use only electrical tools or test equipment that is grounded or triple insulated.
- If possible, never work alone on a live piece of equipment.

SUMMARY

This introduction chapter on electricity will assist you in the study and mastery of the material in the remainder of the text.

Chapter 3

Basic Electrical Circuits

INTRODUCTION

The requirements for a basic electrical circuit are an energy source, an electrical load (energy converter), and conductors to connect them—for example, the incandescent bulb connected to the battery in Figure 3-1. The pictorial of the circuit is shown in (a), and the schematic is shown in (b), where the bulb is shown as a resistor. The battery is the energy source and the bulb is the electrical load. The current in the circuit is determined by the basic law of electrical circuits, called *Ohm's law.* Ohm's law describes the relationship between the three properties in the circuit: current, voltage, and resistance. We use the symbol Ω as a shortcut for ohms.

$$Current = \frac{Voltage}{Resistance} \qquad (3.1)$$

$$I = \frac{E}{R}$$

Where I is the current in amperes, E is the voltage in volts, and R is the resistance in ohms.

Example 1: Find the current in the circuit shown in Figure 3-1.

$$I = \frac{1.5V}{1.5\Omega} = 1 \text{ ampere.}$$

The other relationships of Ohm's law are:

$$Resistance = \frac{Voltage}{Current} \qquad (3.2)$$

or

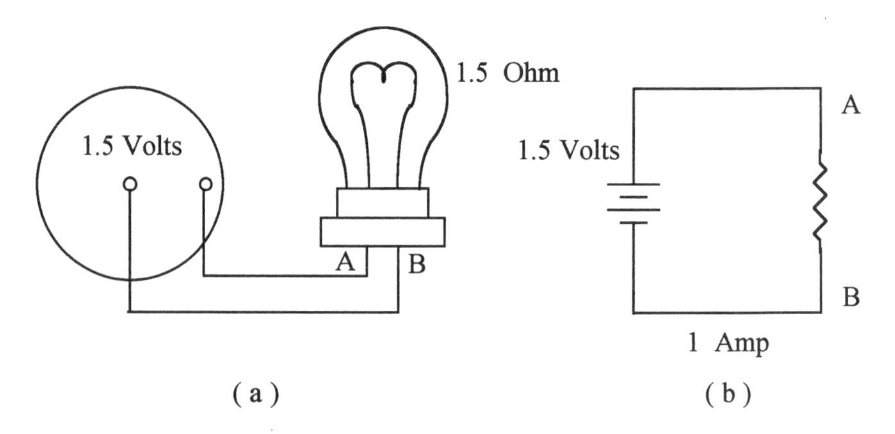

Figure 3-1. An electrical circuit: (a) the pictorial (b) the schematic.

$$R = \frac{E}{I}$$

and

$$E = I \times R \quad (3.3)$$

Note: Voltage is symbolized by the letters E or V.

To summarize these units: current is defined as the movement of electrons (energy) within the circuit, resistance is defined as the opposition or limiting factor to current, and voltage is the electromotive force that drives the current through the circuit.

WIRES AND THEIR ELECTRICAL PROPERTIES

Most conductors are made of copper, although aluminum is also used where light weight and economy are primary factors. Copper is a better conductor than aluminum and should be used whenever possible.

Electrical wire is usually round. However, square, rectangular, or copper strip conductors are sometimes used. The copper on printed-circuit boards is utilized in many circuit applications.

A printed-circuit board is manufactured by plating a copper film on an insulating material. The copper is then etched to form conductors and component pads. The advantages of printed-circuit boards are that they can be manufactured in large quantities, at reasonable cost, and with precision. The consistency of printed-circuit allows them to be assembled by robots.

Conductors may be bare or insulated with materials such as rubber, varnish, asbestos or plastic. Insulation covers the conductor preventing short circuits between conductors and ground. The type of insulation used on a conductors is determined by the environment in which the conductor is to be used. Factors such as moisture, heat, acid, and voltage must be considered when selecting insulation types.

The basic description of round wire is its diameter, which is usually expressed in a gauge number. A gauge number refers to the diameter of the wire, and disregards any insulation. Table 3-1 gives the size, the current capacity (ampacity), and the resistance of the most common wire sizes. The resistance and the ampacity of the wire is based on 1000 feet of length.

Table 3-1. Wire table of common copper wire sizes.

Wire Size	In Conductor or Cable	In Free Air
20	5 A	7 A
18	12 A	15 A
14	15 A	20 A
12	20 A	25 A
10	30 A	40 A
8	45 A	65 A
6	65 A	95 A
4	85 A	125 A
2	115 A	170 A
1	130 A	195 A
0	150 A	230 A
00	175 A	265 A
000	200 A	310 A

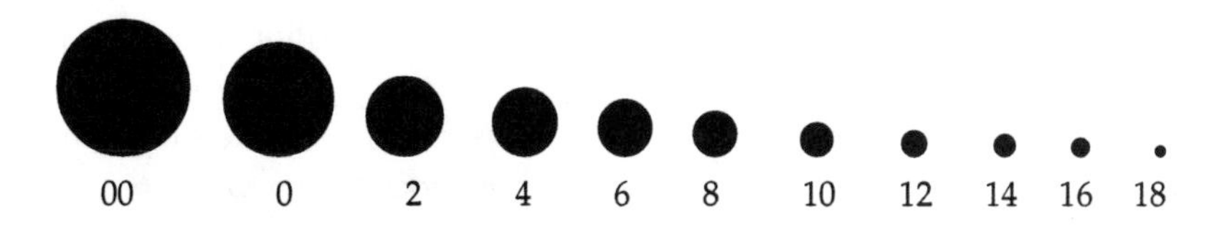

The standard gauge is called the American Wire Gauge (AWG), or the Browne & Sharpe gauge. Figure 3-2 is an example of an AWG gauge tool that is used to measure the diameter and gauge of a wire.

Figure 3-2. An example of an American Wire Gauge.

The resistance of typical copper wire is approximately 10.4 ohms per circular mil foot. A mil foot is a wire with a diameter of 1/1000 of an inch and one foot long. Let us consider, in Figure 3-3, the relation of series and parallel wires, first making a wire larger in diameter by placing two 1 mil wires in parallel, and second, consider the effect of making the wire longer by placing two 1 mil wires in series.

For two 1 mil foot wires placed side by side, in parallel, the resistance would be reduced to 5.2 ohms. On the other hand, two wires placed in series will result in doubling the resistance to 20.8 ohms.

From the foregoing, we can conclude that conductors placed in parallel result in a decrease in resistance. Furthermore, we can also conclude that conductors placed in series result in increased resistance. These are two important properties of series and parallel connections that we will explore later.

In any electrical installation, the conductor size or area must be sufficient to carry the required current without too large a voltage drop

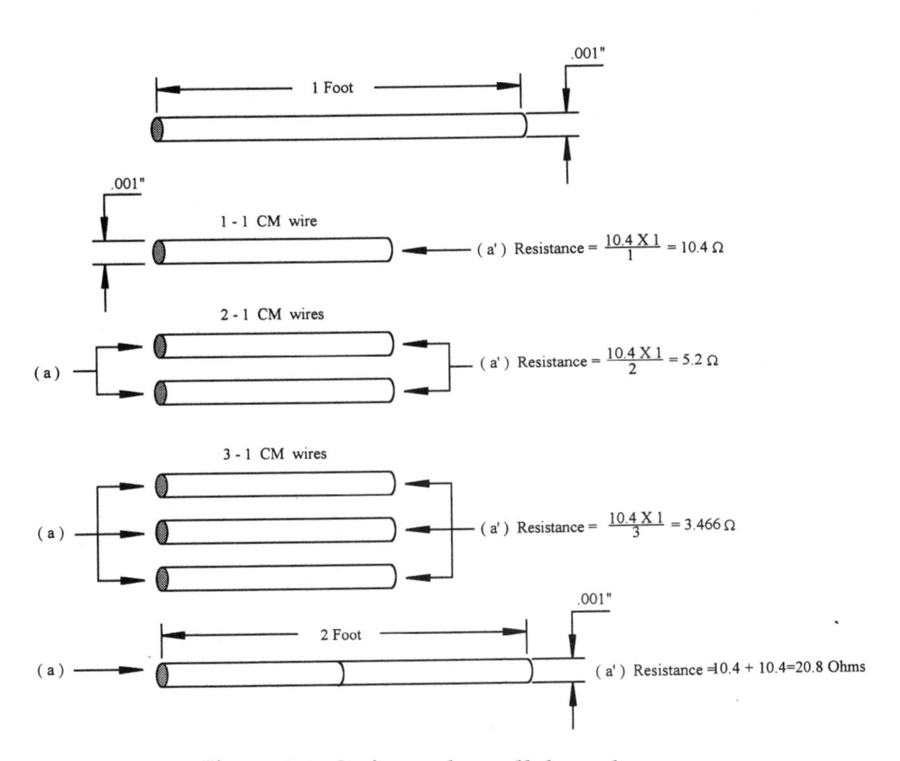

Figure 3-3. Series and parallel conductors.

or excessive heat. The voltage drop in a wire is caused by the current in the conductor and the wire resistance. The smaller the wire diameter, the larger the resistance. A conductor that is too small may have a large resistance which will cause a drop in voltage. This voltage drop takes voltage away from the load, preventing it from operating properly. In an extreme case the resistance of the wire could produce enough heat to melt the insulation and cause a short circuit or fire.

LOW VOLTAGE ELECTRONIC DEVICES

There are four basic electronic components: resistors, capacitors, inductors, and control devices. However, there are thousands of ways to combine these components into circuits.

- Resistors are made of materials that limit current, such as carbon or nicrome wire.

- Capacitors are devices that store electric charge, and oppose a change of voltage.

- Inductors or coils are devices that can store an electric charge, and oppose a change of current.

- Control devices, such as transistors and diodes, control current flow or shape waveforms.

The basic and most often used components are resistors. Resistors are used to divide voltages, or to limit the amount of current in a circuit. Composition resistors, such as illustrated in Figure 3-4, are made in ohmic values and wattage sizes of 1/16, 1/8, 1/4, 1/2, 1, 2 and 4 watts. The wattage rating of a resistor is the maximum power that it can dissipate without overheating. Resistors of greater than 4 watts are made of resistive wire.

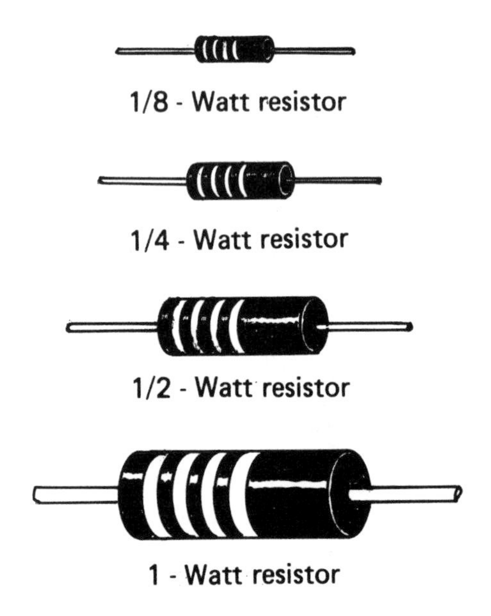

Figure 3-4. Examples of carbon resistors.

The ohms value of composition resistors is written on the resistor or given by the standard *resistor color code.* The resistor color code, and its application are covered in the Addendum at the rear of the text.

Variable resistors that are adjustable are called potentiometers. Figure 3-5 shows an example of a 500 ohm 20 watt potentiometer. This potentiometer can be adjusted from zero to 500 ohms. Applications of potentiometers are the fine adjustment temperature controls found on some electronic-type air-conditioning control circuits.

Resistors, and potentiometers are subject to failure such as burning open or changing their value. These components can be tested with an *ohmmeter.* All these components must be replaced with ones of the same ohmic value and same wattage rating.

BASIC CIRCUIT

Figure 3-6 depicts the most basic electrical circuit. When the switch is closed that circuit has a complete path for current to flow out of the negative terminal of the battery and into the positive terminal. The current through the lamp resistance causes heat and light. When the switch is opened there is no path for current and the light is off.

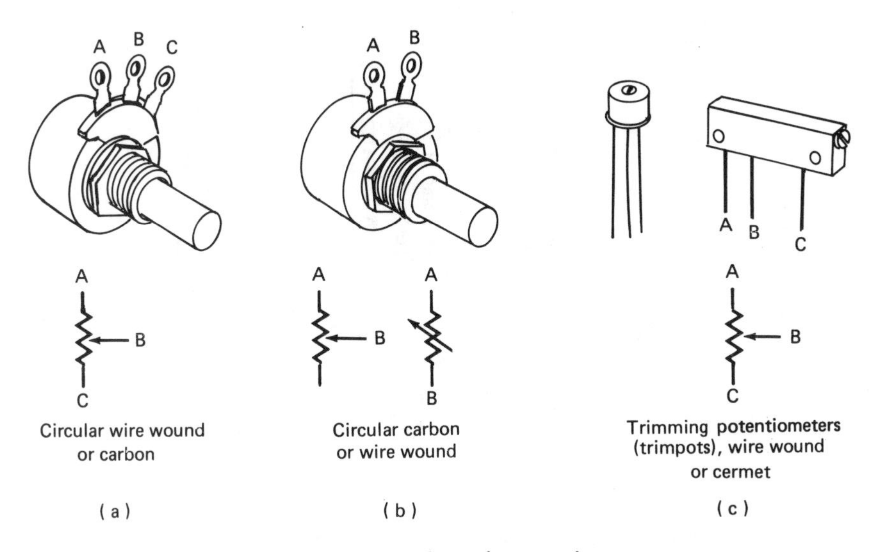

Figure 3-5. Examples of potentiometer.

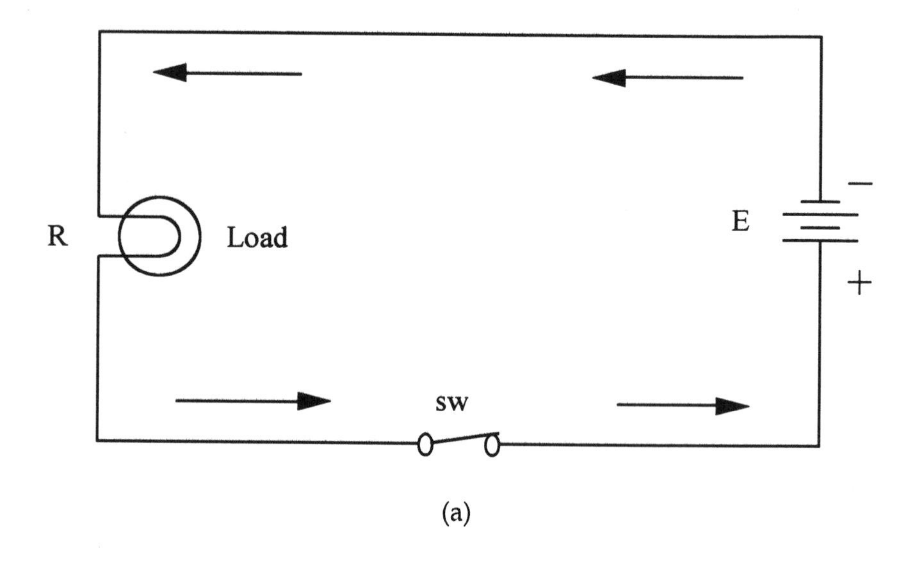

(a)

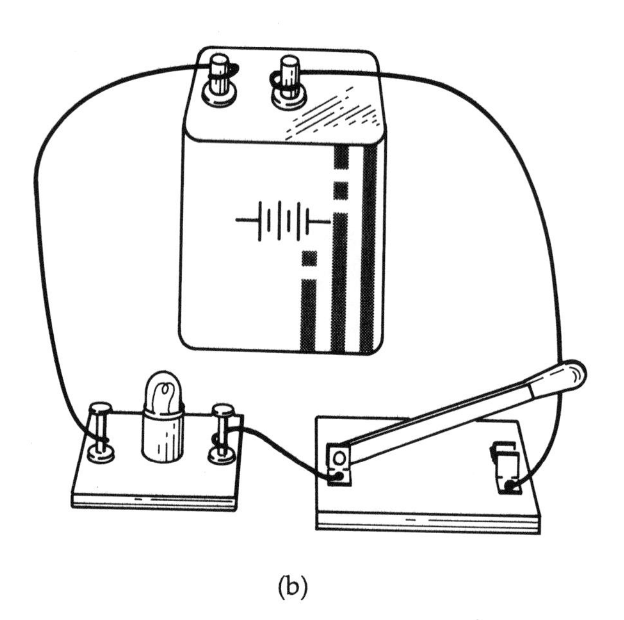

(b)

Figure 3-6. A basic circuit with a lamp, battery and switch.: (a) circuit, (b) pictorial.

Switches are used to turn circuits on and off. There are many types of switches in general use. The most common type is the toggle switch shown in Figure 3-7. This is called a single-pole-single-throw (SPST) switch. The symbol for switch (SW) is shown on Figure 3-7(b). More complex toggle switches are used to switch more than one circuit. For example, a single-pole-triple-throw (SPTT) is three switches in one, and turns on and off three circuits at one time.

There are a number of types of switches and several methods of turning them on and off. Chapter 9 gives a summary of switch types.

A switch must be capable of handling the current and voltage of the circuit.

ELECTRICAL POWER

Electricity is useful because it does work. For example, electricity operates the compressor of an air conditioner, runs the blower motor on

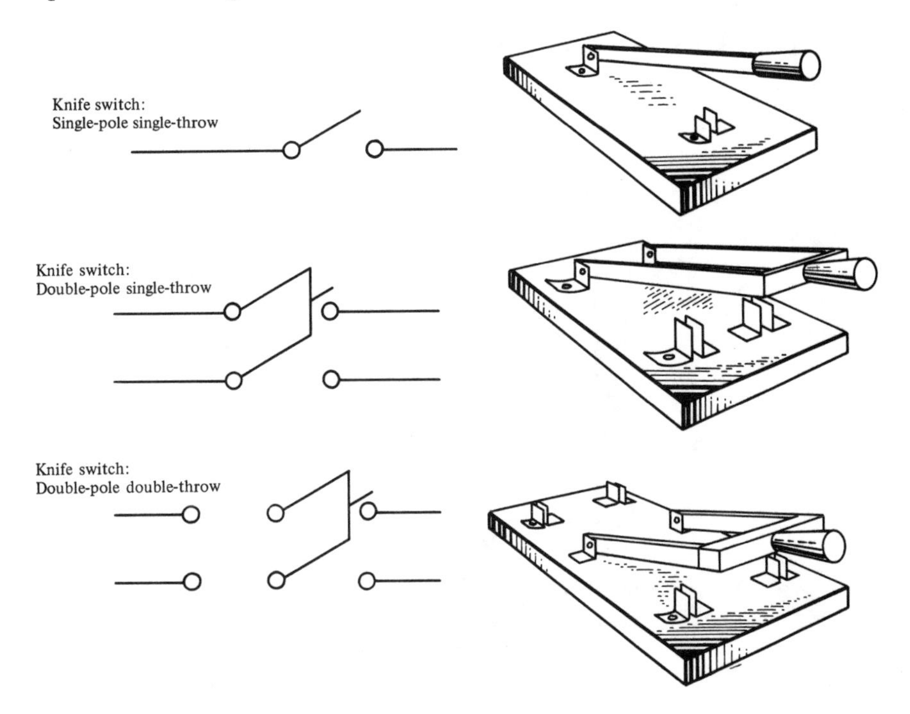

Figure 3-7. Examples of switches.

a furnace, powers a computer and thousands of other devices that we depend upon daily. Work is measured in foot-pounds, as is shown in Figure 3-8. When a one-pound weight is lifted 3 feet above the testing position, three foot-pounds of mechanical work has been accomplished. The weight has gained three foot-pounds of *potential energy*. The amount of work is the same whether it is lifted in one minute or one hour. Time is not a factor.

Power in an electrical circuit is the *time rate* of doing work, and is an important quantity in electrical circuits. Power is the product of current and voltage of a circuit and is given in watts. When one volt moves one ampere past a point in one second one watt of energy is converted.

$$\text{Power} = \text{Current} \times \text{Voltage (Watts)} \quad (3.1)$$

$$P = I \times E \quad \text{(Power Law)}$$

Power in an electrical circuit is related to horse power:

$$1 \text{ Horse Power} = 746 \text{ Watts} \quad (3.2)$$

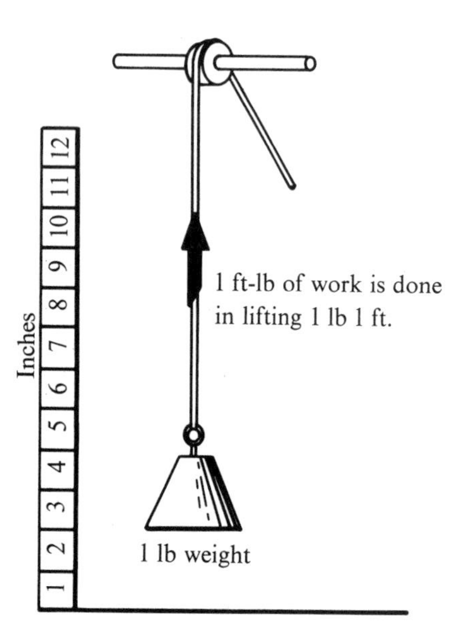

Figure 3-8. Demonstration of measurement of work.

For example: What is the electrical power required to operate a 20 horsepower motor ignoring losses?

$$20 \text{ Hp} \times 746 \text{ Watts/Hp} = 14{,}920 \text{ W or } 14.92 \text{ kilo Watts}$$

CURRENT AND VOLTAGE

The voltage, resistance, current, and power of an electrical circuit are interrelated by Ohm's law and the Power law. The technician must be able to calculate these quantities and measure them with electrical instruments. Figure 3-9 shows an electrical circuit with an ammeter and voltmeter connected to make measurements. The voltmeter must always be connected across the component or line to measure voltage. The ammeter, on a multimeter, must always be connected in the circuit to measure current. The exception is the clamp-on ammeter which is connected

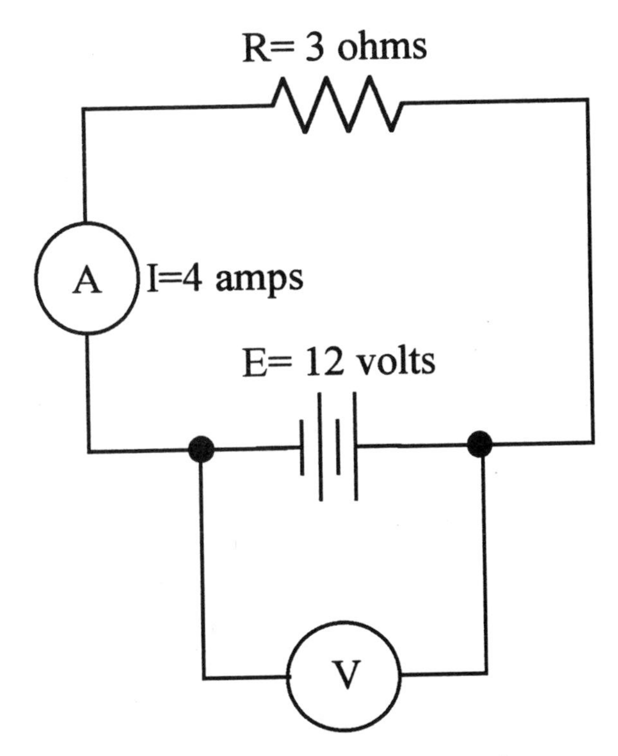

Figure 3-9. A simple circuit with a voltmeter and ammeter connected

around the wire. The ammeter indicates 4 amperes of current and the voltmeter indicates a 12-volt reading. The resistance is found by Ohm's law.

$$R = \frac{E}{I} = \frac{12V}{4A} = 3\Omega$$

Ohm's law can be used to find the current when the voltage and resistance are given.

$$I = \frac{E}{R} = \frac{12V}{3\Omega} = 4A$$

Finally, Ohm's law can be used to find the voltage when the current and resistance are given.

$$E = IR = 4A \times 3\Omega = 12 \text{ V}$$

The form of Ohm's law to use can be determined by the *Ohm's Circle* in Figure 3-10. A finger is placed over the quantity to be solved and the formula remains.

The power dissipated in the circuit can be found three ways.

$$P = 4A \times 12V = 48 \text{ Watts}$$

$$P = E^2/R = \frac{(12)^2}{3\Omega} = \frac{144}{3} = 48W$$

$$P = I^2R = (4)^2 \times 3\Omega = 48W$$

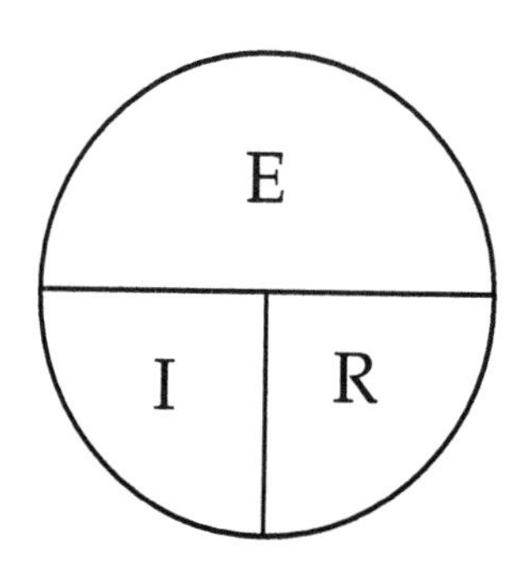

Figure 3-10

SUMMARY

- The current carrying capacity of conductors must not be exceeded.
- The insulation on a conductor must be able to endure any adverse condition present within its environment, such as heat, moisture, or acid.
- The American Wire Gauge or the Browne & Sharpe gauge are used to measure the diameter of wires.
- The ampacity of a wire is its current-carrying capability.
- The unit of resistance is the ohm (Ω), and the symbol is R.
- The ampere is the unit of current, and the symbol is A.
- The volt is the unit of voltage, and the symbols are E and V.
- There are 746 watts in one horsepower.
- Power is the time rate of doing work in an electrical circuit, and the unit of power is the watt.

Chapter 4

Electrical Circuit Analysis

INTRODUCTION

Electricity is an exact science. If we understand electrical theory the action of current and voltage can always be predicted. It is our responsibility to understand electricity and how it will perform in an electrical circuit. We must understand how a circuit is expected to operate before we can determine when it is not functioning correctly. Only then can we locate and repair problems.

Figure 4-1 is a circuit diagram of a window-type air conditioner. The circuit shows the interrelationship of switches, relay contacts, the thermostat breaker, the air circulating fan, capacitors, and the compressor motor windings. At this point in our study we cannot comprehend the purpose of each component nor the sequence of operation. We must first study basic circuit action and the electrical characteristics of each component. When we return to this circuit in Chapter 14 you will be able to comprehend the function of each component, their relationship to each other and the sequence of operation.

The technician must be able to read and interpret electrical wiring diagrams so that he or she may test circuits, and locate and repair circuit problems. Wiring schematics and diagrams are the technicians' road maps for understanding the operation of circuits and systems.

SERIES CIRCUITS

We will begin the study of electrical circuits with series circuits, because this is the simplest of electrical circuits. For example, the series-connected lamps in Figure 4-2 form a basic circuit. The lamps are connected end-to-end, so that current from the voltage source moves through each lamp. The supply voltage is dropped across the total cir-

Figure 4-1. Wiring diagram of a window-type air conditioner.

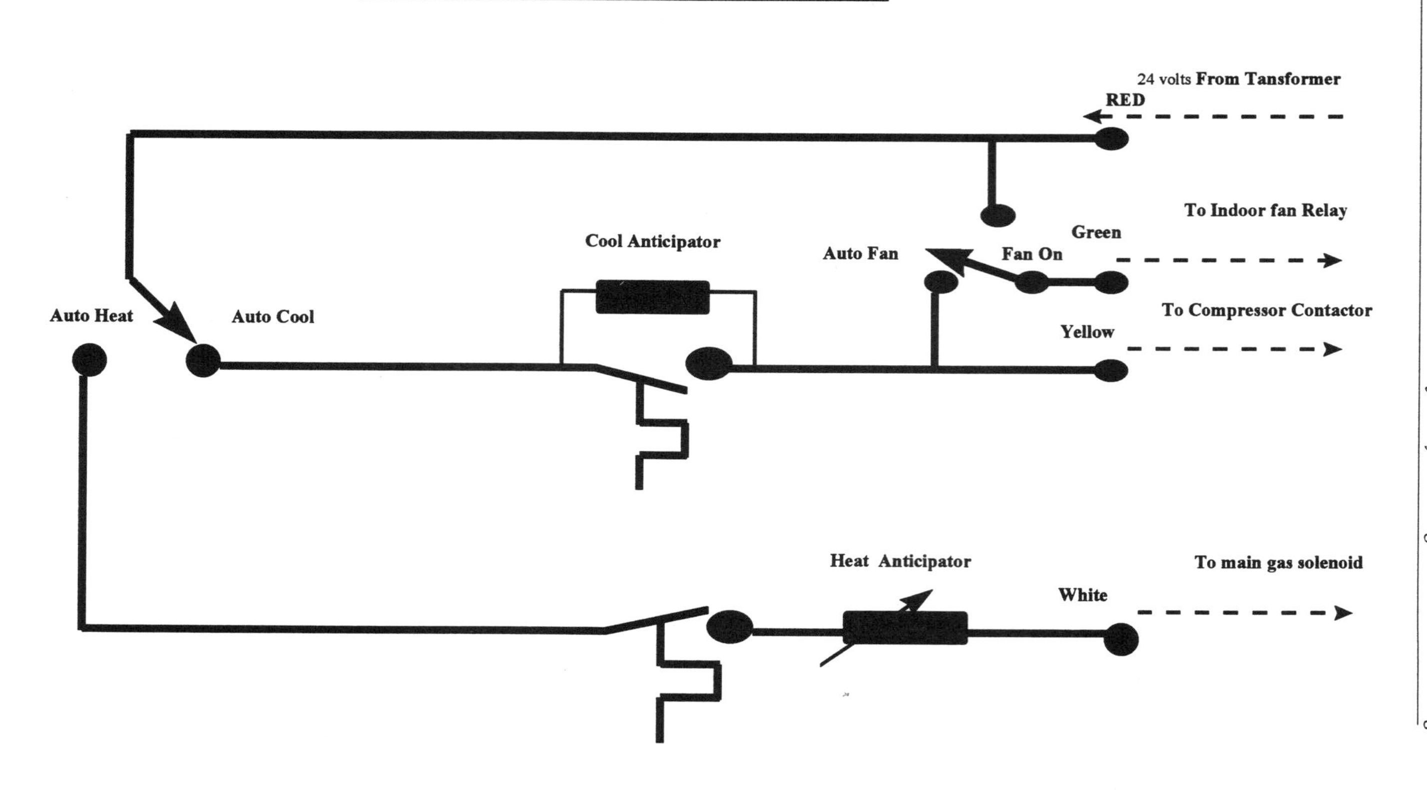

cuit. The voltage drops across each of the lamps will depend upon their resistance. The sum of their voltage drops will equal the value of the supply voltage. If the resistance of each bulb is the same, the source voltage will be dropped equally across each bulb as shown in Figure 4-3.

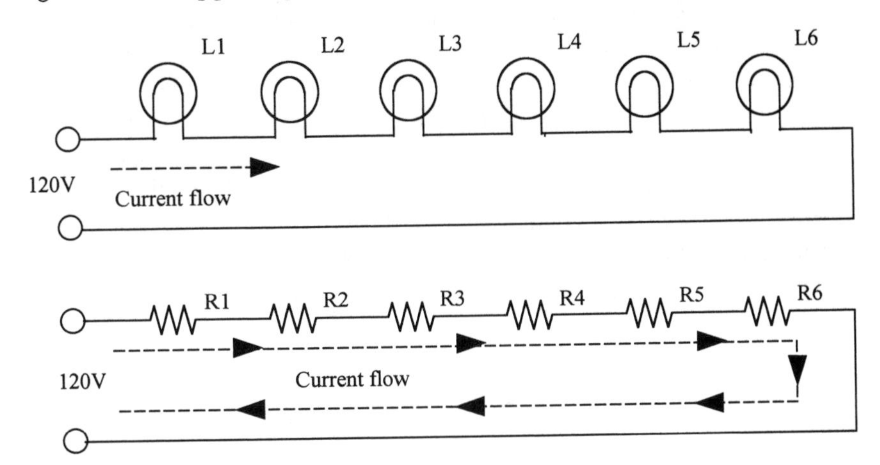

Figure 4-2. A series Christmas tree lamp circuit: (a) appearance of the circuit, (b) schematic diagram, (c) equivalent circuit.

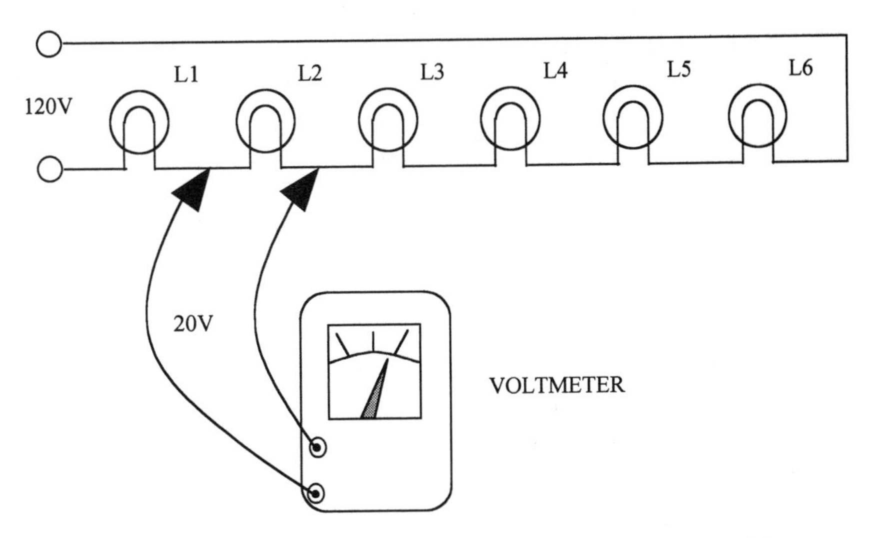

Figure 4-3. The voltmeter measures the voltage across L2.

When the string is connected to a voltage source the current enters one end of the lamp string and leaves the other end If the lamp filament of one bulb in the series string burns out, an *open circuit* occurs. The effect is as if the switch in Figure 4-4 were opened. No current would flow and no voltage would be dropped across the bulbs. However, if we connect a voltmeter across the switch we would read the entire 120 volts. This is because the voltmeter completes the current path allowing some small current to flow. The extremely high resistance of the voltmeter, in comparison to that of the lamps, causes all the voltage to be dropped across the voltmeter. Note that the internal resistance of a DVM is 10 to 11 meg ohms. If the switch is closed, the voltmeter across the switch would read zero and the voltmeter across L2 would read 120 volts. When a component burns out in a series circuit we say there is an *open-circuit fault*.

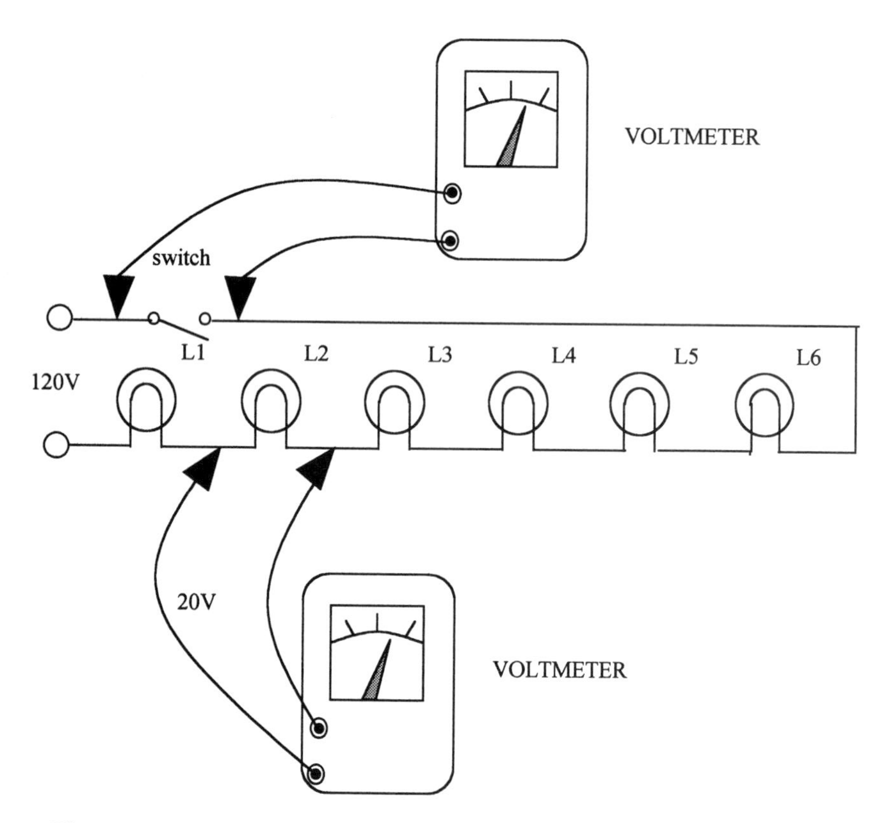

Figure 4-4. Demonstration of open-switch action in a series circuit.

Next, let us consider what happens when a *short-circuit fault* occurs. Figure 4-5 shows a short circuit in a series circuit. A short is often caused by frayed or worn insulation on wires. When the bare wires in Figure 4-5 touch together, a short circuit occurs. The short circuit has a very low resistance in respect to the lamp filaments, effectively eliminating them from the circuit. This results in a very large current through L1 and L2 with the entire 120 volts being dropped across them. There will be 60 volts across each, probably resulting in one or both being burned open. On the other hand, the other four lamps will not be damaged.

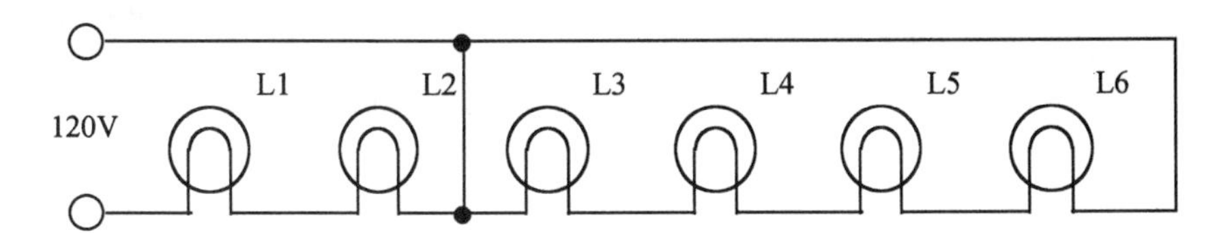

Figure 4-5. Short in a series circuit.

Suppose one of the six lamps in the circuit in Figure 4-3 is open, and we wish to find the faulty one. With the switch closed and voltage applied, the open lamp can be found by *bridge testing*. Bridge testing is accomplished by placing one lead of the voltmeter on the input and moving down the line of series components. When the opened component is passed, the voltmeter will read the source voltage. Another method would be to place the leads across each of the components in turn. The voltmeter would again read source voltage when it was bridged across the faulty lamp. *The switch must be closed and the power on to make either of these tests.* This is because the very high resistance of the voltmeter completes the circuit, dropping all the voltage across itself.

SERIES CIRCUIT SUMMARY

We may summarize these facts as follows:

$$I_T = I_1 + I_2 + I_3 \tag{4.1}$$

Equation 4.1 states that the current is the same in all parts of a series circuit.

$$E_T = V_1 + V_2 + V_3 \quad (4.2)$$

Equation 4.2 states that the sum of all the voltages around any closed loop must equal zero when we consider the sign of the voltage drop.

$$R_T = R_1 + R + R_3 \quad (4.3)$$

Equation 4.3 states that the total resistance is equal to the sum of all resistors.

The circuit in Figure 4-2(b) represents the schematic diagram of the lamp circuit. The schematic diagram is utilized for simplicity of drawing and understanding of circuit action. Figure 4-2(c) represents the equivalent circuit of the lamp string. The equivalent circuit is utilized for simplicity of analysis of the circuit. Suppose that each of the lamps has a resistance of 7.5 ohms. We find the total resistance of the string with formula 4.3.

$$R_T = R_1 + R_2 + R_3 + R_4 + R_5 + R_6$$

$$= 7.5 + 7.5 + 7.5 + 7.5 + 7.5 + 7.5 = 45 \text{ ohms}$$

We find the total current of the circuit from Formula 4.4.

$$I_T = \frac{E_T}{R_T} \quad (4.4)$$

$$I_T = 120\text{V}/45 \text{ ohms} = 2.67 \text{ A}$$

The voltage across each lamp is:

$$V = I \times R = 2 \text{ A} \times 7.5 \text{ ohms} = 15 \text{ volts.}$$

PARALLEL CIRCUITS

Parallel circuits are widely used in electrical and electronic equipment. Almost all residential lights and appliances are connected in parallel, for example, the Christmas-tree lights in Figure 4-6(b). There are

120 volts across each lamp. Also, each lamp provides a complete path for current flow. An equivalent circuit for the parallel string is shown in Figure 4-6(b).

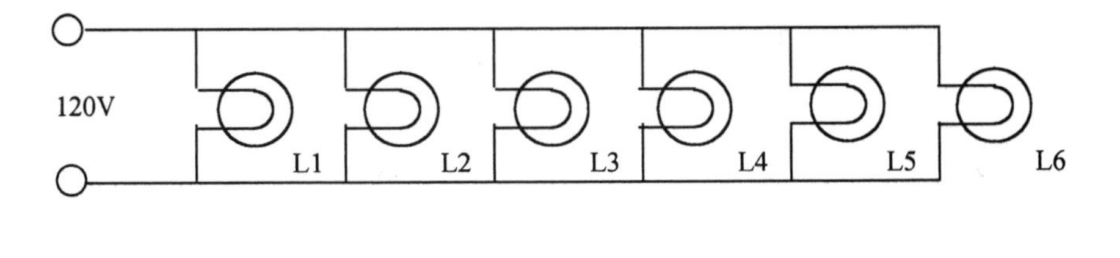

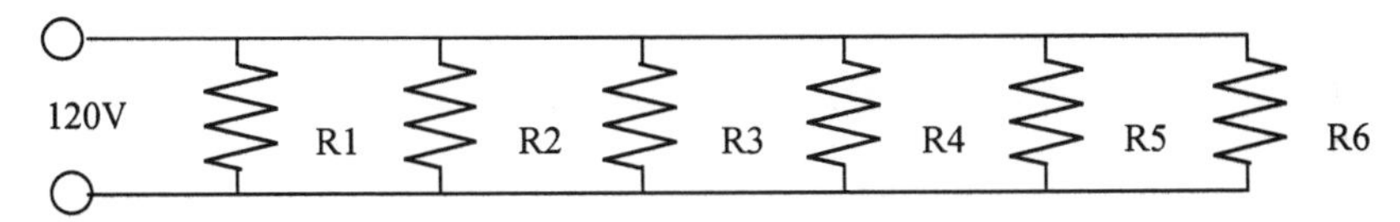

Figure 4-6. Parallel string of Christmas lights: (a) appearance of string, (b) schematic diagram, (c) equivalent circuit.

In the equivalent circuit each lamp in the string is represented by a resistor. Since all bulbs are the same the resistors are of equal value. Opening or removing one of lamps will result in a reduction of total current but will have no effect on the other lamps. This is because the voltage across, the current through, and the power dissipated for each of the lamps will not change. Figure 4-7 illustrates the current within the parallel circuit. The current in each path adds to the next from right to left. For example, if the resistors were all 120 ohms, a current of 1 ampere would flow through each.

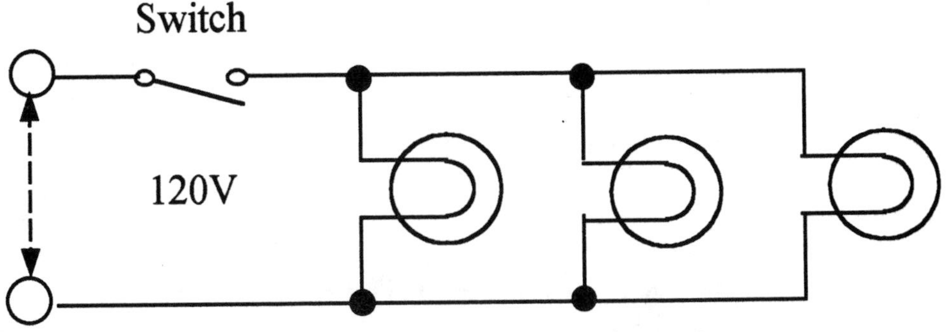

Figure 4-7. Current flow path in a parallel circuit.

A total current of 6 amperes would flow from and return to the source. The current would decrease by 1 ampere at each junction. The rule for current flow is that *the current into a junction equals the current out of the junction.* For example, current I2 would equal:

$$I_1 = I_T + I_1 + I_2 + I_3 + I_4 + I_5 + I_6$$

$$I_1 = 6A - 5A = 1A$$

A parallel circuit may have a basic switch to remove the voltage from all components as shown in Figure 4-8. When the switch is closed, all are lighted, and when the switch is open all lamps are dark. On the other hand, a parallel circuit may have a switch for each load; for example, the light circuits and the appliance circuits in a residence where each load is independent of all others. The main circuit breaker would be in series with all loads and would be used to remove all power.

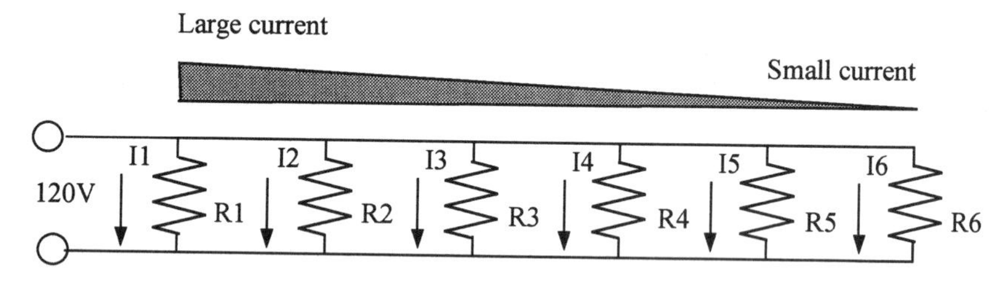

Figure 4-8. A parallel bank of lamps controlled by a switch.

PARALLEL CIRCUIT SUMMARY

- The voltages across all components in a parallel circuit are equal.
- The current in each parallel component depends on Ohm's law:
- $I = \frac{E}{R}$
- The current into a junction is equal to the current out of the junction in a parallel circuit.

- The total resistance of a parallel circuit is less than the resistance of the smallest resistance and is formulated:

$$\frac{I}{R_t} = \frac{I}{R_1} + \frac{I}{R_2} + \frac{I}{R_3}$$

The total power dissipated in a parallel circuit is the sum of the power dissipated in each component, and is also the product of total current times total voltage.

SERIES-PARALLEL CIRCUITS

A series-parallel circuit consists of one or more series circuits combined with one or more parallel circuits. A simple example of a series-parallel circuit is shown in Figure 4-9. This arrangement is for dimming lights. An adjustable resistor, or rheostat, is connected in series with a group of parallel lights. As the contact is moved to the right more resistance is added to the circuit and circuit current decreases, thereby decreasing the voltage across the lights. In turn, they become dimmer. Of course, moving the contact to the left decreases the resistance in the circuit causing more current to flow and the light to brighten. To calculate the currents and voltages in a series-parallel circuit requires the application of most of the electrical circuit laws.

Example 4.1: Find the current in each resistor, the voltage across, and the current through each resistor in the circuit in Figure 4-10.

1. Find the total parallel resistance of $R_2 \parallel R_3$

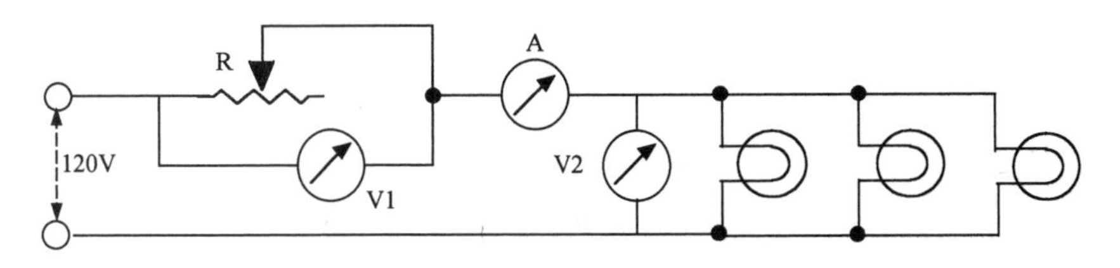

Figure 4-9. A light-dimming arrangement forming a series-parallel circuit.

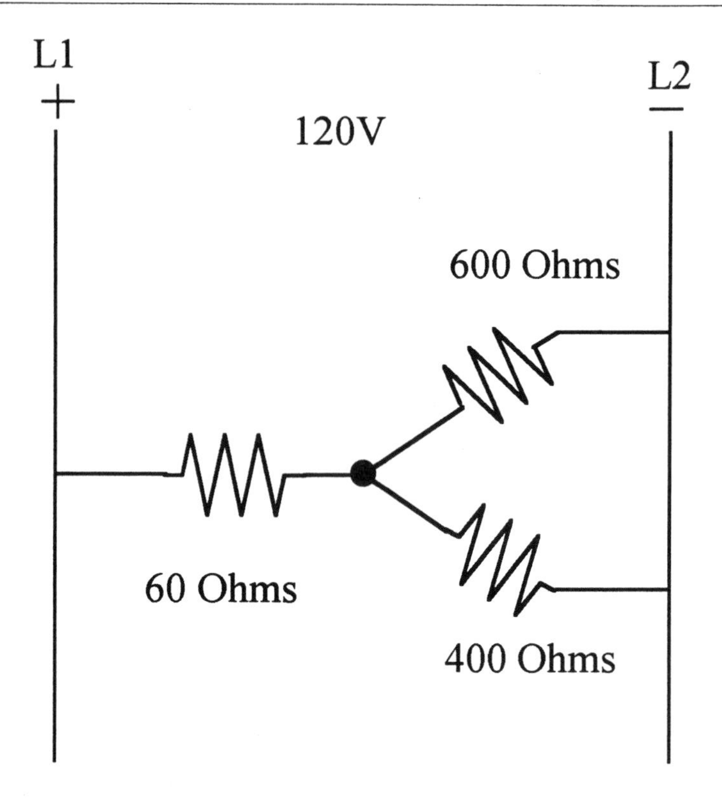

Figure 4-10. A series-parallel circuit.

$$R_P = \frac{R_2 \times R_3}{R_2 + R_3}$$

$$R_P = \frac{R_2 \times R_3}{R_2 + R_3} = \frac{600 \times 400}{600 + 400} = \frac{2400}{1000} = 240\ ohms$$

2. Add R_1 to R_P

   $R_T = R_1 + R_P = 60 + 240 = 300\ ohms$

3. Find the total current by Ohm's law;

   $$I_R = \frac{E_T}{R_T} = \frac{120V}{300\ ohms} = 0.4A$$

4. The voltage across R_1 is:

$$E_1 = I_1 \times R_1 = 0.4A \times 60\ ohms = 24\text{V}$$

5. Since the voltage around a circuit must equal the applied voltage, we deduce that the voltage across the parallel network is:

$$120V - 24V = 96V$$

6. or we might calculate the parallel voltage:

$$V_P = I_T \times R_p = 0.4A \times 240\Omega = 96V$$

$$V_P = V_2 = V_3 = 96V$$

7. The current through V_2 and V_3 is found by Ohm's law.

$$I_2 = \frac{96V}{600\Omega} = 0.16A$$

$$I_3 = \frac{96V}{400\Omega} = 0.24A$$

Their sum is the total current 0.4 A.

Series-parallel circuits can be very complex. However, with a systematic step-by-step approach and the application of circuit laws they can be evaluated.

CIRCUIT TROUBLESHOOTING

Series Circuits

Faults can be found in a series circuit by the use of a voltmeter, an ohmmeter, or replacement of each component in turn.

To summarize series circuit troubleshooting:

- An open switch, wire connection, or any component in the circuit will result in no circuit current.

- Source voltage applied, the switch closed, and only one component open will result in a voltage reading of the source voltage across the defective component.

- An open circuit can be found by bridge testing.

- A defective component in a series circuit can be found with an ohmmeter by removing the source voltage and measuring across each component. An open component will have a resistance of infinity.

- A short circuit in a series circuit results in increased current and usually a blown fuse or circuit breaker.

- The total resistance of a series circuit is the sum of each resistor.

$$R_T = R_1 + R_2 + R_3 \ldots$$

- The total voltage in a series circuit is the sum of all the voltage drops.

$$E_T = E_1 + E_2 + E_3$$

- The total power dissipated in a series circuit is the sum of the power dissipated by each component or the product of the total current and total voltage.

$$P_T = P_1 + P_2 + P_3 \ldots$$

Parallel Circuits

Parallel circuit faults are somewhat easier to locate than those in series circuits. Usually, only one of the parallel components or devices is inoperable. A faulty component can be isolated, repaired, or replaced.

SUMMARY

1. The ohmmeter is used to make resistance readings and continuity tests, but only with the power off.

2. The voltmeter is always connected across the source or the component for which the voltage is to be measured.

3. An ammeter, other than the clamp-on ammeter, is always connected in series with the circuit.

Chapter 5

AC Circuits

ALTERNATING VOLTAGE AND CURRENT

Air conditioning, heating and refrigeration systems operate on alternating current and voltage (AC). The standard voltages are 120 volts, 240 volts or 440 volts. The frequency of these voltages is 60 cycles per second. Cycles per second is identified as Hertz or Hz. Figure 5-1 illustrates a single cycle of a 120-V 60Hz waveform. The voltage waveform is developed by an AC generator that may be powered by water, gas or coal. The output waveform a generator develops is called a sine wave. The waveform repeats itself 60 times each second, and passes from a maximum positive value to a maximum negative value. Obviously, the sine voltage can be defined by its peak value or its peak-to-peak value. However, these two points are of little interest in environmental control systems.

We are interested in the amount of work a voltage performs, or the effect that a voltage has on a circuit. Therefore, AC voltages and currents are measured and identified by their *effective value.*

In the circuit with a resistor (Figure 5-2), the AC voltage and current perform work during both their positive and the negative cycles.

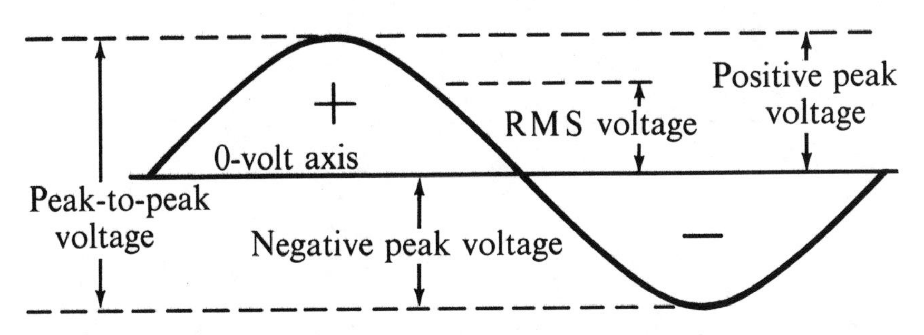

Figure 5-1. Details of an AC sine wave.

Therefore, the power waveform can be considered to perform as illustrated in Figure 5-2, where both halves of the waveform are positive.

$$P = I \times E \text{ Watts} \tag{5.1}$$

$$P = I^2R \tag{5.2}$$

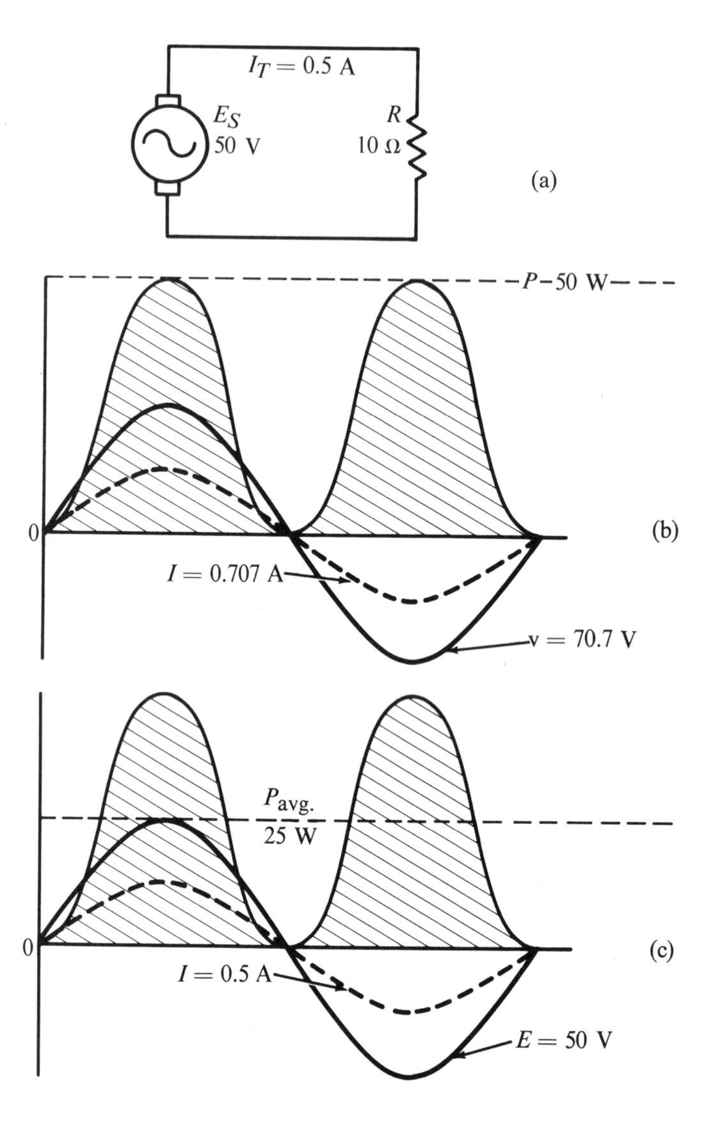

Figure 5-2. Graphic representation of current, voltage and power in an AC circuit.

During the positive cycle both current and voltage are positive and power is positive. Also, during the negative cycle both voltage and current are negative and their product is positive as shown in Figure 5-2. The current and voltage waveforms in a resistor are in phase.

$$P = -I \times -E = +P \text{ Watts} \tag{5.3}$$

These conditions prevail only in a purely resistive circuit. The product of current and voltage is the formula for power. This formula is accurate only in circuits with only resistance. The correct formula is:

$$P = I \times E \text{ cosine}\Theta \tag{5.4}$$

In a resistive circuit the phase angle between current and voltage is zero degrees, and the cosine of zero degrees is 1. Therefore, the cosine Θ being one and can be dropped.

Example 5-1. What is the power dissipated in a 100 Ω resistor connected across a 120-V source?

$$I = \frac{E}{R} = \frac{120V}{100\Omega} = 1.2A$$

$$P = I \times E = 1.2A \times 120V = 144 \text{ Watts}$$

or

$$P = I \times E \text{ cosine}\Theta = 1.2 \times 120 \times 1 = 144 \text{ Watts}$$

INDUCTORS IN AC CIRCUITS

Inductance is the property of a coil that opposes a change of current in a circuit. The property of the inductor causes the voltage to lead the current by 90 degrees in an AC circuit. In a pure inductor, one without resistance, this causes an unusual effect on the power as illustrated in Figure 5-3. Since the current and voltage are always out of phase by 90 degrees their product, the power, switches from positive to negative each half cycle. This results in both a positive and negative power wave.

The effect is that the inductor takes power from the source and returns the power back to the source. The final result is zero power dissipation. Since there is both current and voltage in an inductive circuit, but no power, we must define the condition by another term. The term that is used is *Volts Amperes Reactance (VARS).*

Example 5.2: What is the power in a pure inductor, having an inductive (X_L) of 100 Ω and connected across a 120V AC source?

$$I = \frac{E}{X_L} = \frac{120V}{100\Omega} = 1.2A$$

where X_L = iInductive reactance

$$P = I \times E = 1.2A \times 120V = 144\ VARS$$

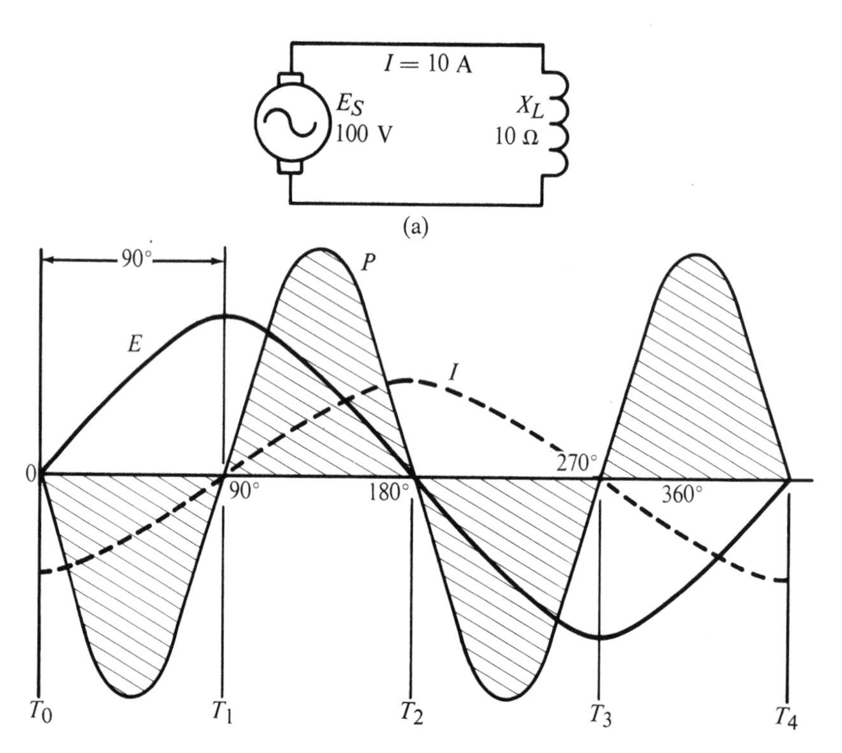

Figure 5-3. Phase relations in an inductive circuit: (a) circuit, (b) voltage, current, and power relations.

A wattmeter connected in the circuit in this example would read zero watts because no power is dissipated, whereas a VAR meter would read the product of current and voltage. The power factor (P.F.) of a circuit can be found from the ratio of apparent power to true power.

$$P.F. = \frac{True\ Power}{Apparent\ Power} \tag{5.4}$$

CAPACITORS IN AC CIRCUITS

Capacitance is the property of a capacitor that opposes any change of voltage in a circuit. As circuit voltage changes, a capacitor charges and discharges, slowing down voltage changes across its plates. The current in a capacitor leads the voltage across the capacitor by 90 degrees, as shown in Figure 5-4. The product of current and voltage is a power wave that is similar to that of the inductor. Again, the power dissipated is zero and the current and voltage product is defined as VARS.

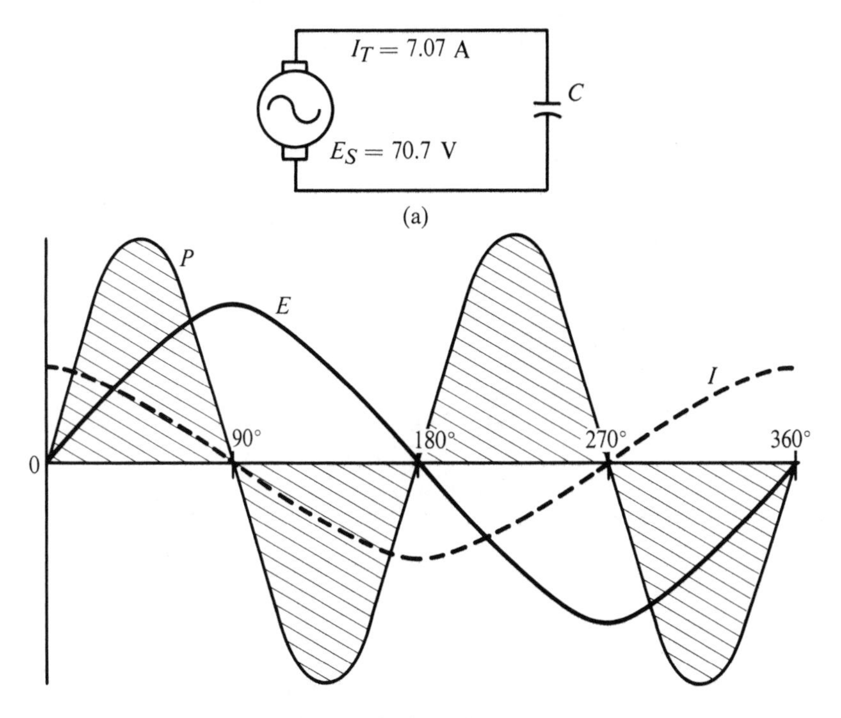

Figure 5-4. Phase relations in a capacitor circuit.

Example 5-3: What is the resulting power dissipated when a capacitor with a reactance X_c of 100 Ω is connected across a 120-V source?

$$I = \frac{E}{X_c} = \frac{120V}{100\Omega} = 1.2A$$

$$P = I \times E = 1.2A \times 120V = 144VARS$$

MEASURING AND CORRECTING POWER FACTOR

Figure 5-5 illustrates the current, voltage, and power relations in an RL circuit. The power dissipated by the resistor is positive and the reactive power of the inductor is negative. The actual power that would be

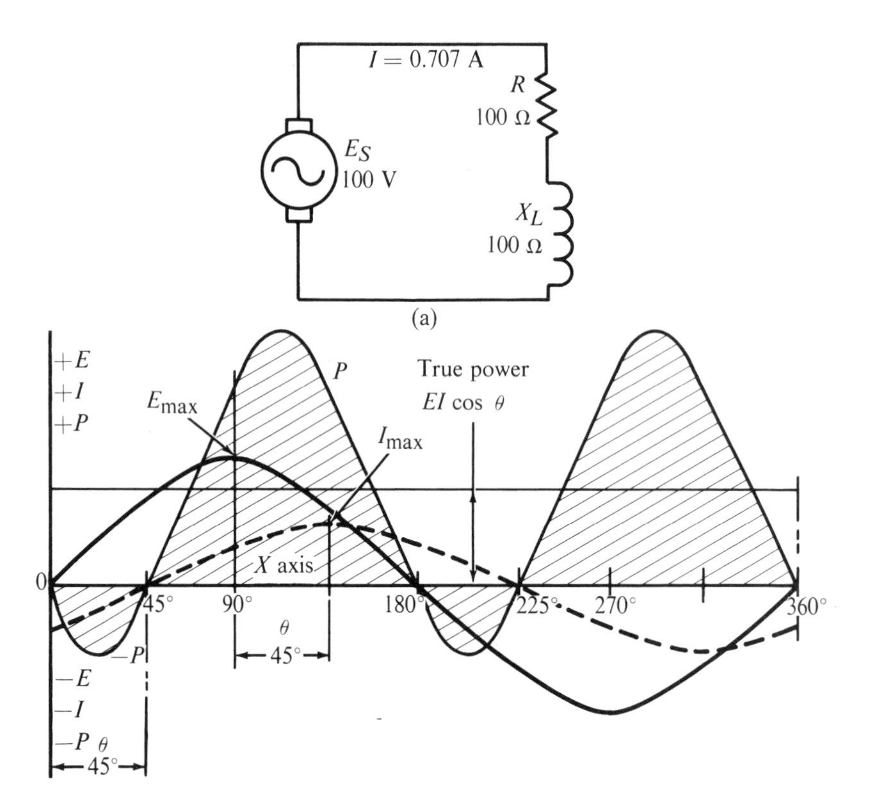

Figure 5-5. The relationship of current, voltage and power in an AC circuit.

measured by a wattmeter is the power of the resistor only.

The power factor of a circuit can be calculated, but one must measure the current and power of the circuit.

Example 5 4: Find the P.F. of a motor when a clamp-on ammeter measures 10 A, the voltage is 440 V_j, and a wattmeter measures a true power of 2500 W.

$$P = I \times E$$

$$Apparent\ Power = I \times E = 10A \times 440V = 4400VARS$$

$$P.F. = \frac{2500W}{4400VARS} = 0.568$$

The phase angle between the current and voltage in the motor is the angle whose cosine is 0.568. Calculate this angle by placing 0.568 in the calculator, and pressing INV and COS to read 55.9 degrees.

The power factor of a motor or the entire facility can be measured with a P.F. meter, sometimes called a VAR meter.

When a large number of inductive motors are used in a facility, the system will operate with a low power factor. This means that the true power measured by the kilowatt-hour meter is much less than the power delivered by the power company. This, of course, does not make for a happy supplier.

The result is that the utility may charge an additional fee for the power measured, or insist in a correction of the power factor. The P.F. can be corrected by placing the correct value of capacitor in the line. The correct value can be calculated from measured values. However, most utility companies will supply a chart or booklet that details the method of selection of the capacitor and the proper method of installation.

SUMMARY

- The current and voltage in a resistive circuit are in phase.
- Power dissipated in resistance is the only true power.
- Capacitors cause the current to lead the voltage in a circuit.
- Inductors cause the current to lag the voltage in a circuit

- The relationship between the current and voltage in inductors and capacitors can be remembered by the rule; **ELI** the **ICE** man, where L and C represent inductors and capacitors; E and I represent voltage and current.

- Power in a capacitor or inductor is apparent power in VARS.

- Power measured by a wattmeter is true power—that dissipated by resistance in a motor or other circuit component.

- The power factor can be calculated from measured and known circuit values.

- The power factor can be measured with a P.F. meter.

- A low power factor can be corrected by the addition of the correct value of capacitance.

Chapter 6

Capacitors and Reactance

INTRODUCTION

Capacitors are also called *condensers,* although the name capacitor is preferred by technicians. Capacitors are widely used in air conditioning, refrigeration, and heating equipment to affect the flow of alternating current.

A capacitor is a device that stores and discharges electricity, and has the ability to shift the phase of an AC current. There are several types of capacitors; however, all types consist of metal plates separated by an insulating material called a dielectric. The simplest capacitor is two metal plates separated by air as the dielectric.

We will observe how capacitors work with an example. However, we must first know some facts about capacitors.

- Any capacitor has a certain amount of capacitance.
- A capacitor offers opposition to a current flow.
- The less the capacitance of a capacitor, the greater is its opposition to alternating current.
- A capacitor has less opposition to a high frequency than a low frequency.
- A capacitor has the ability to shift the phase of a current.
- A capacitor has a maximum working voltage (WV).

Example 6.1

We may demonstrate the circuit action of a capacitor by the following example in Figure 6-1. An electric light bulb is connected in series with a capacitor and a source voltage. When the circuit is connected to a DC voltage, the bulb does not light; but, when an AC voltage is applied to the circuit the lamp glows. This experiment shows that a capacitor blocks the flow of DC current, and allows the passage of AC current. If a larger capacitor were placed in the circuit the lamp would glow brighter, showing that a larger capacitor offer less resistance to AC current.

The opposition that a capacitor offers to a current flow is called *reactance*. A capacitor has capacitive reactance, in much the same way that an inductor has inductive reactance. Capacitive reactance is measured in ohms just as inductive reactance is measured in ohms.

Although capacitive reactance can be compared to resistance, we find that capacitive reactance is not the same as resistance. For example, if we replaced the capacitor with a resistor in the circuit in Figure 6-1, the lamp would glow in each circuit. Perhaps we can understand capacitor action on AC current by the drawings in Figure 6-2. A capacitor is connected in a series circuit with a switch, a capacitor, and an ammeter. Before the switch is closed there is no voltage across the capacitor, and the capacitor is uncharged. In Figure 6-2(b), when the switch is closed, there is a momentary surge of electron out of the upper plate and into the lower plate. This quickly charges the capacitor to the battery voltage, and no current flows after the momentary surge.

In Figure 6-2(c) the battery has been replaced with a conductor and the switch is open. The capacitor maintains a charge equal to that of the battery and there is no current flow. Finally, the switch is closed in Figure 6-2(d) and there is a momentary surge of electrons out of the lower plate into the upper plate and the charge on the plates is equalized. The capacitor is now discharged and the voltage across the plates is zero.

CAPACITORS IN AC CIRCUITS

Let us now observe an example that shows how AC current flows in a capacitor. A capacitor is connected in series with a galvanometer and a double throw switch to a battery, as shown in Figure 6-3. A galvanometer is a center reading ammeter that can read current in either direction. When the switch is in the up position, the right-hand plate of the

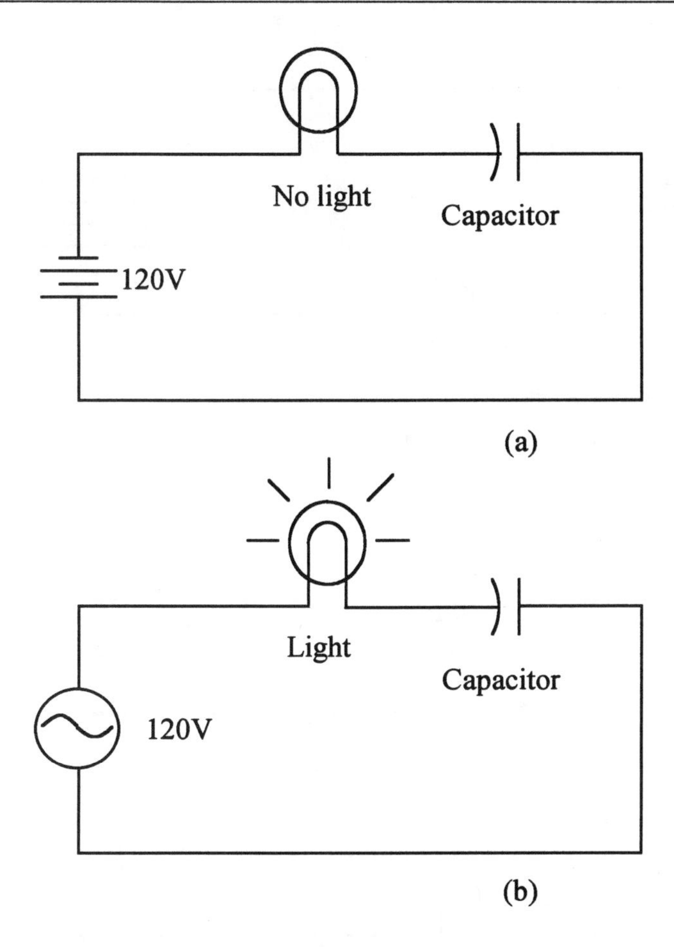

Figure 6-1. Capacitor action: (a) current is blocked with DC voltage, (b) current flows with AC voltage.

capacitor is negatively charged as electrons are collected. On the other hand, when the switch is in the down position, the right-hand plate of the capacitor is charged positively and the left-hand plate is charged negatively. When the switch is switched up and down rapidly, the galvanometer deflects from left to right, as long as the switch action is continued. If the switch were replaced with an AC source, the current would flow back and forth with each change of the sine wave.

The unit of capacitance of a capacitor is the *farad*. A capacitor that has a capacitance of one farad takes a charge of 1 coulomb when connected to a 1-volt source.

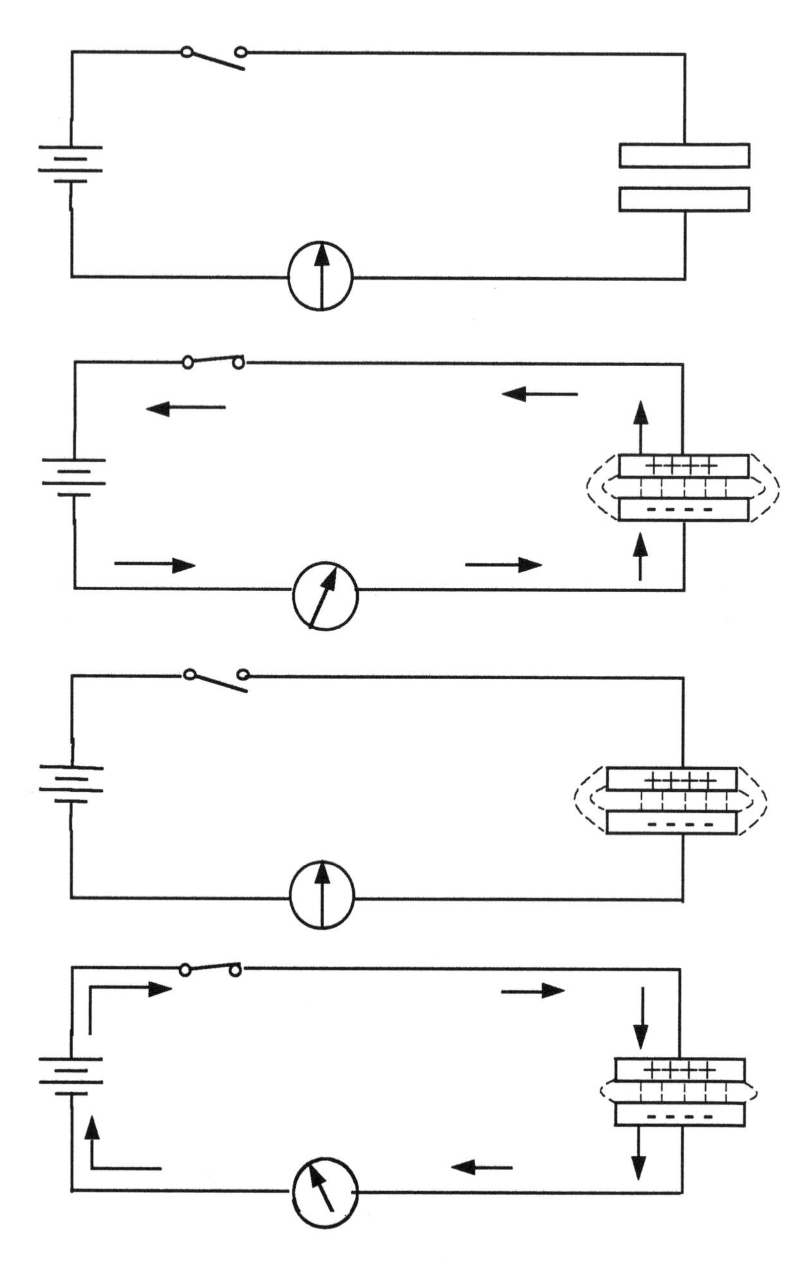

Figure 6-2. Capacitor action: (a) charging circuit, switch open, (b) charging circuit, switch closed, (c) discharging circuit, switch open, (d) discharging circuit, switch closed.

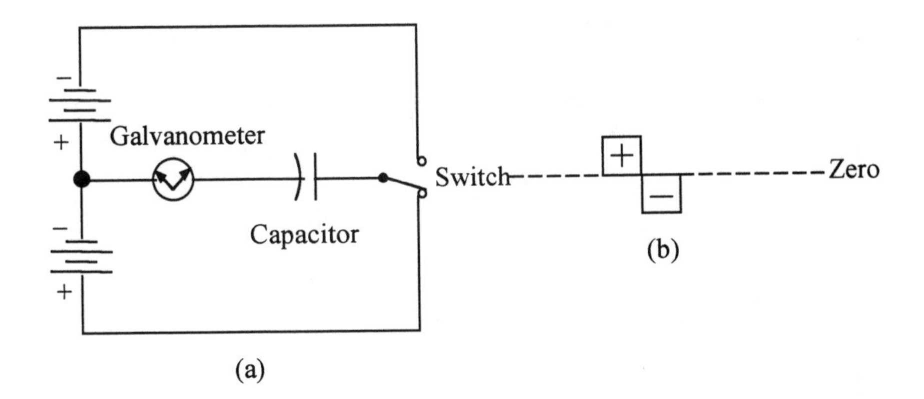

Figure 6-3. An example of how AC current flows in a capacitor: (a) circuit, (b) AC current graph.

- One coulomb is equal to 6.25×10^{18} electrons. Note that micro farad is written µF, nano farad is written nF, and Pico farad is written pF. A µF is 1/1,000,000 of a farad, a nF is 1/1,000,000,000 of a farad, and a pF is 1/1,000,000,000,000 of a farad. Capacitors are manufactured in values of 2 pF up to 10,000 µF.

- Each capacitor has a maximum voltage value called WV. For example a motor capacitor may be rated at 200 µF and 500 WV.

The opposition that a capacitor offers to AC current is called *reactance.* The symbol for capacitive reactance is Xc. The formula for capacitance reactance is:

$$X_c = \frac{1}{2\Pi fC} \tag{6.1}$$

The current in an AC circuit with a capacitor is calculated by Ohm's law.

$$I = \frac{E}{X_c}$$

For example, determine the reactance and current in the circuit in Figure 6-4. The reactance is:

$$X_c = \frac{1}{6.28 \times 60 \times 500 \times 10^{-6}} = 5.3\ ohms$$

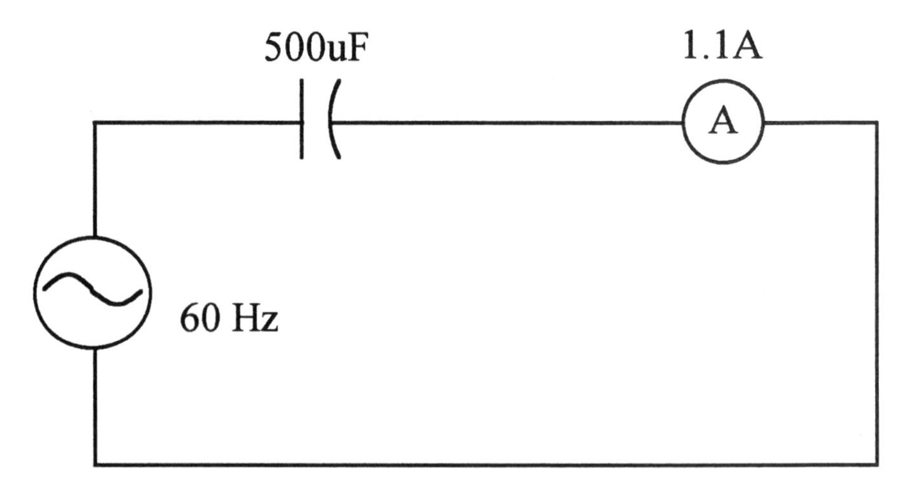

Figure 6-4. A capacitor connected in an AC circuit.

The current is:

$$I = E/X_C = 120/5.3\Omega = 22.6A$$

Many types of capacitors are used in electrical and electronic circuits. Motor and starting capacitors are constructed of sheets of tin foil separated by wax paper insulation, as shown in Figure 6-5. Connecting wires, called pigtails, extend from the foil sheets through the ends of the plastic case or *encapsulation* as shown in Figure 6-5(b). These are electrolytic capacitors.

There are two types of electrolytic capacitors; *polarized and nonpolarized* Most electrolytic capacitors are polarized and made to be used in only DC circuits as one end must always be to the negative voltage and the other to the positive voltage. These capacitors have either the negative or the positive end marked. Always check electrolytic capacitors for the polarization identification. Installing a polarized electrolytic capacitor backwards in a circuit will result in the capacitor getting hot and exploding. Capacitors may be placed in parallel or in series to obtain different values or to obtain a higher voltage rating. When capacitors are connected in parallel their values are added to obtain the total capacitance.

$$C_T = C_1 + C_2 + C_3 \ldots \tag{6.3}$$

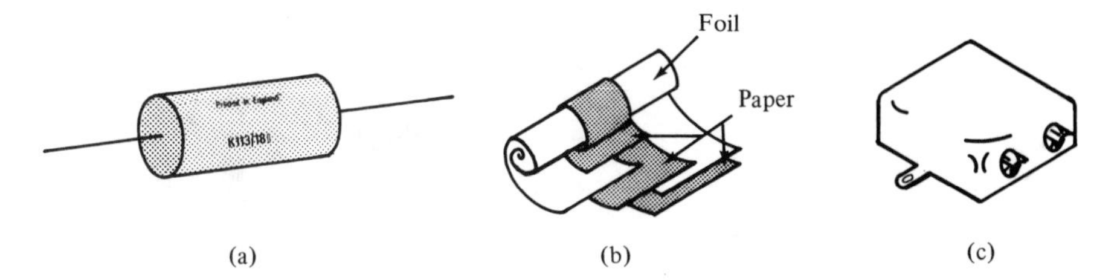

Figure 6-5. Construction details of a capacitors.

The total breakdown voltage of capacitors in parallel is that of the lowest working voltage rating. For example, if three 30 μF capacitors, one with 100 WV, one with 50 WV, and one with 200 WV were placed in parallel, it would result in a 90 μF capacitor at 50 WV.

When capacitors are connected in series, the result is a smaller capacitor and an increase in working voltage.

$$\frac{1}{C_r} = \frac{1}{C_2} + \frac{1}{C_2} + \frac{1}{C_3} \cdots \tag{6.4}$$

For example, if three 30 μF capacitors with 100 WV each were placed in series, the result would be a 10 μF capacitor with 300 WV.

Variable capacitors and adjustable capacitors are used in many electronic circuits, for example, the tuning capacitor in your automobile radio and in air-conditioning and heating controllers. Variable capacitors are used to obtain a precise capacitance value in control circuits. Their value seldom exceeds a few hundred pF. However, some variable capacitors can have working voltages in the thousands of volts.

The two capacitors used in motor circuits are the *start capacitor* and the *run capacitor*. Start capacitors, Figure 6-6 (a), are used to produce an added magnetic torque to assist the motor in overcoming the stationary inertia of the system during starting. These capacitors are removed from the circuit when the motor has reached running speed.

Run capacitors, Figure 6-6(b), remain in the motor winding circuit after starting. The physical distinctions between the two types of capacitors are: both leads of starting capacitor are insulated from the case, and one lead of the run capacitor is connected to the case. Both these capacitors are usually non-polarized electrolytic types.

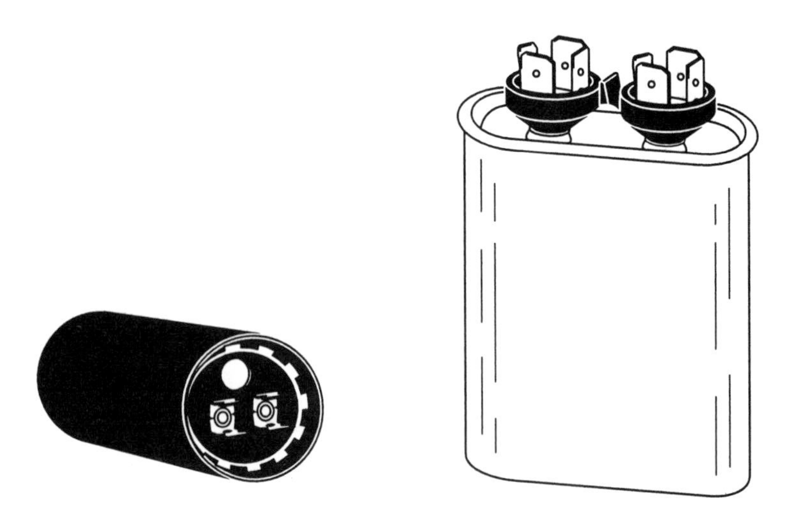

Figure 6-6. Examples of motor capacitors: (a) start capacitor, (b) run capacitor.

SAFETY PRECAUTIONS WHEN WORKING WITH CAPACITORS

There are some important safety precautions that must always be remembered when working with large capacitors.

- A capacitor may hold its charge for many hours. If a high-voltage capacitor is not discharged by short-circuiting its terminals, it may act as a "booby trap" for the unwary technician. Accidental contact with the terminals can result in severe shock, or even death.
- If an electrolytic capacitor is operated above its maximum working voltage, it will "leak" excessively. The leakage current will cause the capacitor to heat and often explode. This precaution applies to both polarized and non-polarized capacitors.
- If a polarized capacitor is connected incorrectly in a DC circuit in the incorrect polarity, it will heat up, and explode.
- If a polarized electrolytic capacitor is connected into an AC circuit, it will heat up, and may explode.

TROUBLESHOOTING AND TESTING CAPACITORS

Most capacitor problems will be the result of the capacitor developing an internal open or an internal short. Both problems can be tested with an ohmmeter. To accomplished this:

1. The VOM should be on the R × 10 or R × 100 scale.
2. One terminal of the capacitor must be isolated from the circuit.
3. Place a VOM across the terminals of the capacitor.
4. The meter pointer should read near zero and slowly rise to near infinity.
5. Reverse the meter leads and the pointer should read less than zero and again rise to near infinity.

An internal short will cause a very low reading and the pointer will not rise. An open will result in the meter reading infinity from the start of the test. Figure 6-7 illustrates the capacitor test. Be aware that a DVM will give inclusive results that cannot be relied upon.

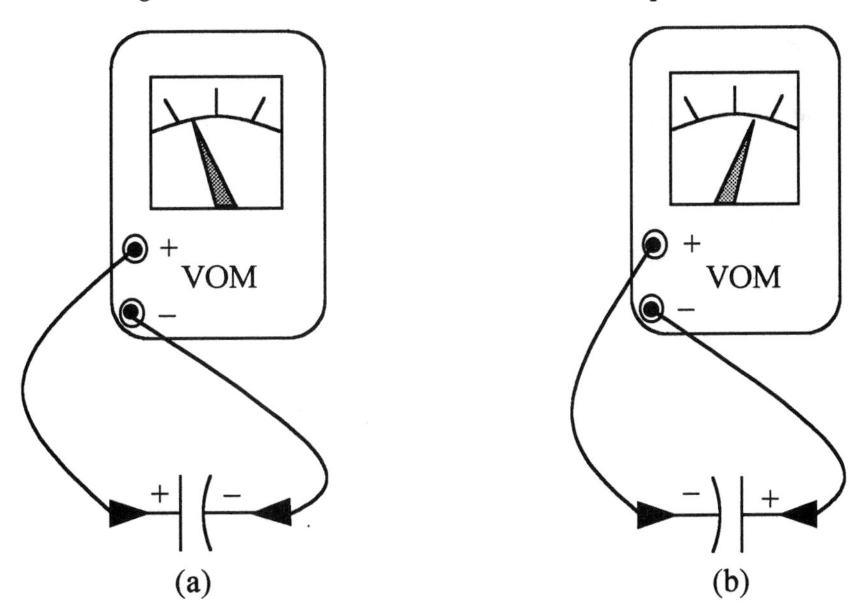

Figure 6-7. Testing a capacitor with an ohmmeter.

SUMMARY

- Capacitors are used in motor circuits to cause a phase shift of winding current.
- Starting capacitors are used to give the motor added starting torque.
- Run capacitors shift the phase in one of the run windings, producing the effect of a two-phase motor.
- Capacitors are subject to internal shorts between terminals.
- Capacitors are subject to internal leakage between the plates.
- As capacitors age they may lose capacitance. The capacitance of a capacitor can only be tested with a capacitance tester.
- A replacement capacitor must be the same type with the same capacitance and working voltage.

QUESTIONS

1. What is another name for capacitors?
2. How is a capacitor tested for an internal short?
3. What is the difference between a run capacitor and a start capacitor?
4. What is meant by the working voltage (WV) of a capacitor?
5. What is the effect of an open run capacitor?

Chapter 7

Magnetism and Inductance

PRINCIPLES OF MAGNETISM

Magnetism has been known of since ancient times. Chinese sailors are said to have used natural magnets, called lode stones, to navigate. When the lode stone was suspended on a string or floated on a cork in water it would point north and south. The end pointing northward is called a north pole and the end pointing south is called a south pole. Today we know that each magnet has a north pole and a south pole and *unlike poles attract and like poles repel.* This means that a magnet acts as a compass because of the weak magnetic field inside the earth's core, which has a south pole at the earth's north pole and a north pole at the earth's south pole. Any magnet is surrounded by a magnetic field. The earth's field is illustrated in Figure 7-1. The lines of the field around a magnet are called *flux lines.* Flux lines are invisible. However, we can prove that they are present, as we observe in Figure 7-2 where the compass needle is deflected by the permanent magnet. As the compass is moved around the magnet, the field of the needle will rotate to align with the field of the magnet.

The important facts for us to demonstrate are from the example :

- The north pole of the compass needle is attracted to the south pole of the magnet.
- The south pole of the compass is attracted to the north pole of the magnet.
- The north pole of the compass is repelled by the north pole of the magnet.

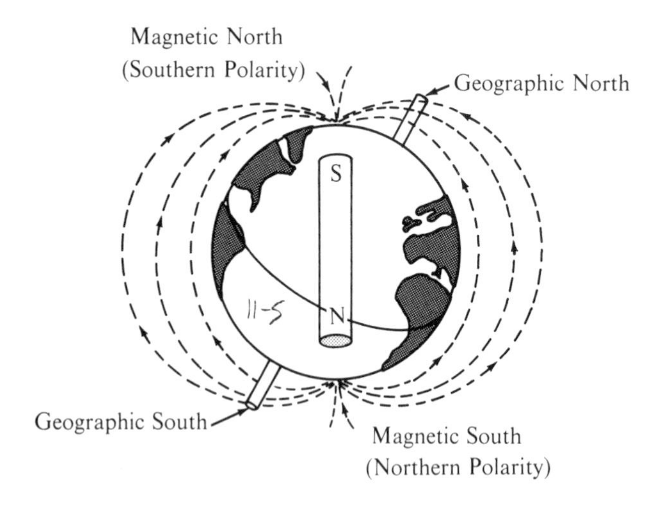

Figure 7-1. The earth acts as a weak magnet.

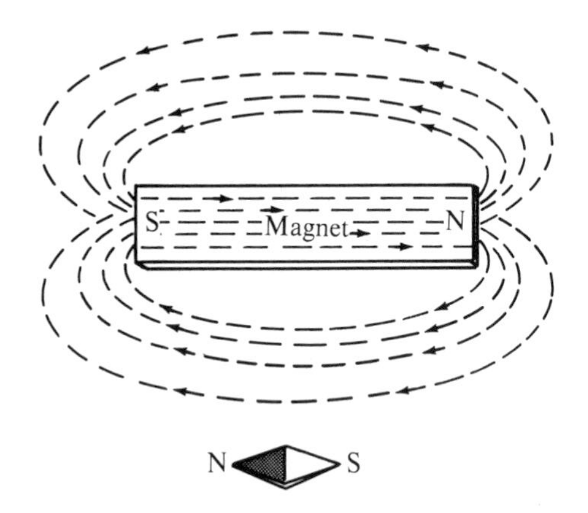

Figure 7-2. The field of a magnet can be explored with a compass.

- The south pole of the compass is repelled by the south pole of the magnet.

We may draw the a representation of the lines of force about a magnet by placing a sheet of paper or plastic over the magnet and sprinkling iron filings on the sheet as shown in Figure 7-3. The iron filings will align with the invisible flux lines. This is because iron is attracted to the lines of force. Each small particle of iron has a magnet induced into it from the flux lines. This induced magnet has its poles opposite to the

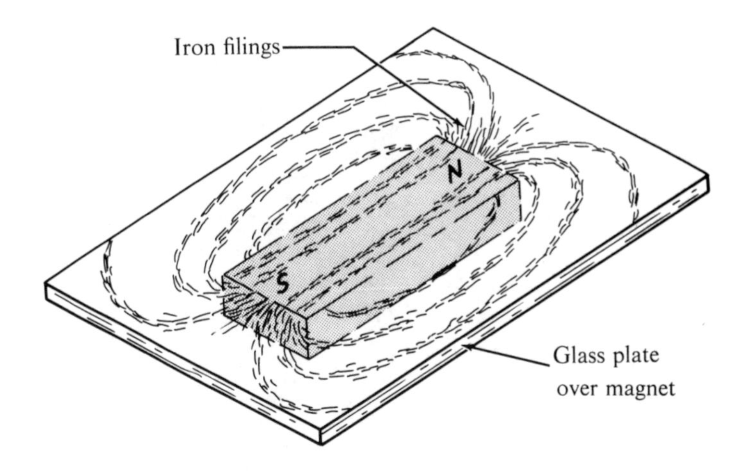

Figure 7-3. Mapping a magnetic field with iron filings.

permanent magnet. Therefore, each filing is attracted to the other. Important facts about magnetic lines of force are:

- Lines of force are considered to come out the north pole and enter the south pole.
- A line is continuous both inside and outside.
- Lines of force never cross each other.

The magnetic field becomes weaker as we go farther from the magnet, but there is no point at which the field disappears completely. When a permanent magnet is formed the molecules within the iron line up in the same direction as shown in Figure 7-4. A permanent magnet may be formed by stroking an iron or steel material with a permanent magnetic as shown. For example, you may magnetize a screwdriver by stroking the tip with a permanent magnet. Each stroke must be in the same direction.

A permanent magnetic may be developed by applying a current to a coil of wire. Let us consider how two magnetic fields act upon each other. The two basic conditions are:

- If unlike forces are facing each other, the magnetic lines of force crowd together, and the two magnets attract each other.

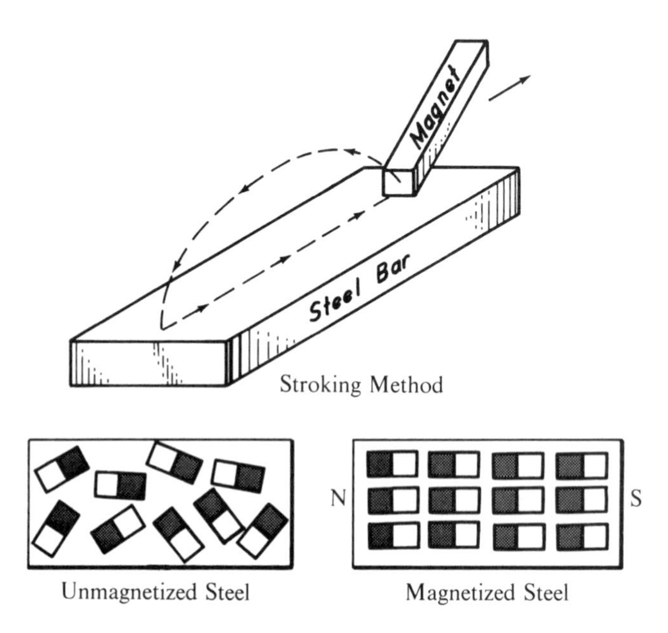

Figure 7-4. Representation of molecular magnets within a steel bar.

- On the other hand, the magnetic lines of force are pushed out from between the poles, and the magnets repel each other.

This brings us to the two laws of magnetic fields:
1. Like poles repel.
2. Unlike poles attract.

In other words a north pole will attract a south pole, a south pole will repel a south pole and a north pole will repel a north pole. We will also find that a south pole cannot exist without a north pole.

MAGNETISM AND ELECTRICITY

There is a close relationship between electricity and magnetism. When a magnetic compass is brought near a current-carrying wire, the compass needle is deflected, as shown in Figure 7-5. The compass needle will position itself across the wire, and will reverse directions if the current in the wire is reversed.

If we map the magnetic field of a current-carrying wire, we will find that the lines of force encircle the wire, as seen in Figure 7-6. The *left-hand rule* is used to determine the direction of the electron flow and the magnetic field. If you clasp the conductor in your left hand with the thumb pointing in the direction of electron flow, your fingers will indicate the direction of flux lines. However, if we form the wire into a spiral or a helix coil, as shown in Figure 7-7, the lines around each wire combine to form a kind of magnet called an *electromagnet.*

It differs from a permanent magnet in that a magnetic field is produced only when a current is flowing in the coil. The left-hand rule for a coil is applied as shown in Figure 7-8. The fingers point in the direction of the electron flow in the coil and the thumb points in the direction of the north pole of the induced magnet. The direction of the magnetic field can be tested with a magnetic compass. When a soft iron core is placed in the coil as shown in Figure 7-9, the magnitude and the strength of the field is greatly increased, because the soft iron has far less resistance to the flux lines than air.

Electromagnets are used in many electrical devices such as transformers, relays, contactors and motors. One example, that we might note here is the relay shown in Figure 7-10. The flow of current through the

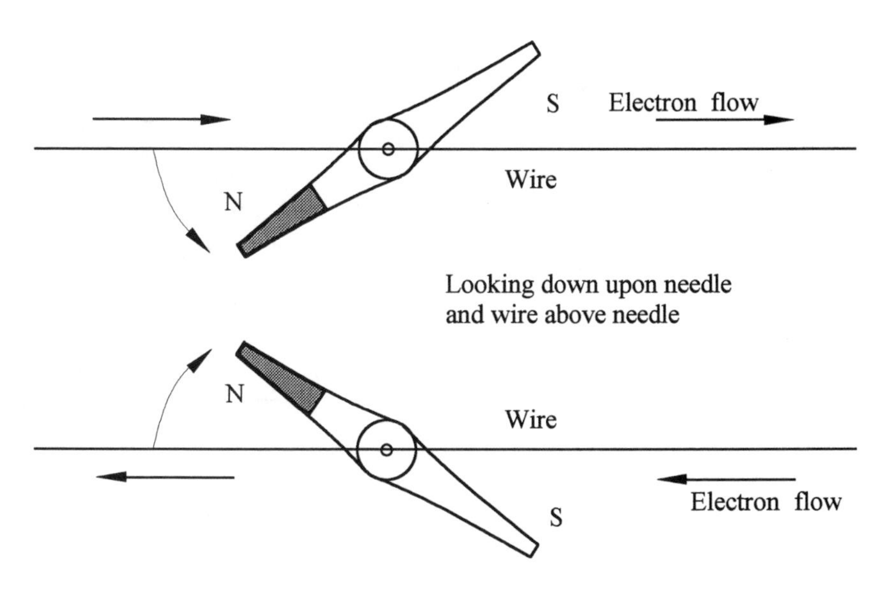

Figure 7-5. A current-carrying wire will deflect a compass.

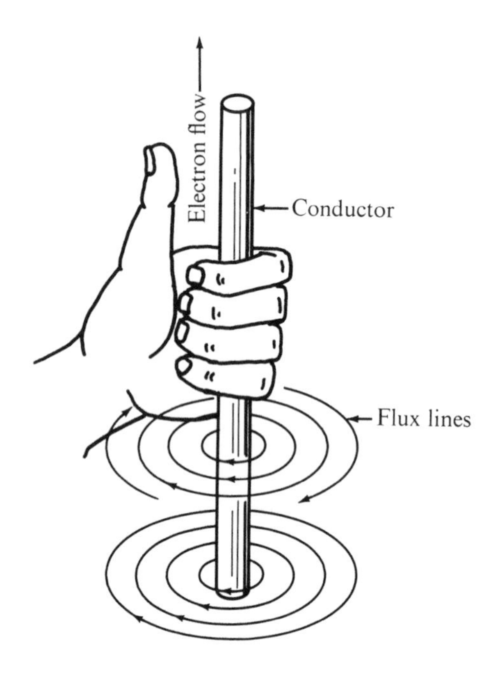

Figure 7-6. Lines of force encircle a wire.

Figure 7-7. Magnetic field around an air-core coil.

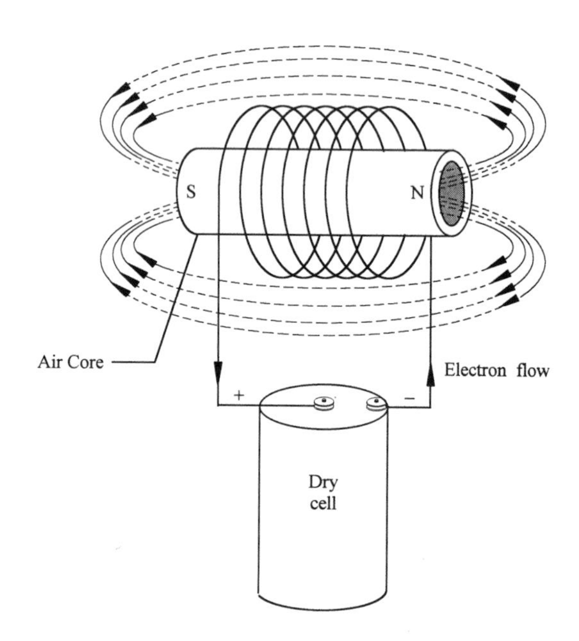

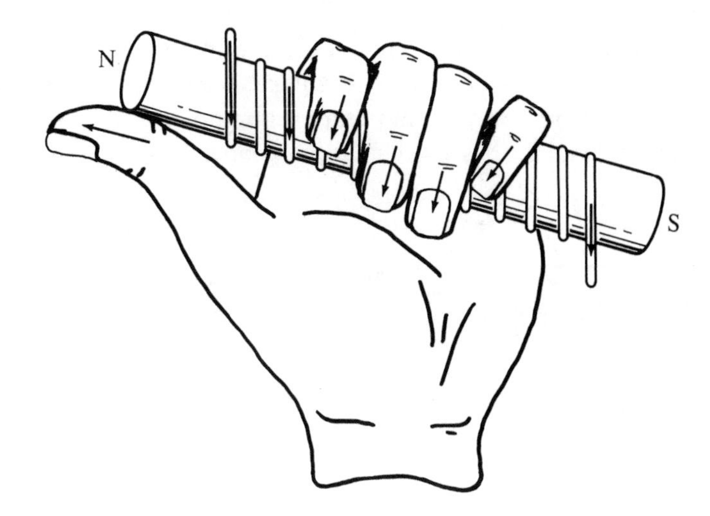

Figure 7-8. The method of finding the magnetic polarity of a coil by the left-hand rule.

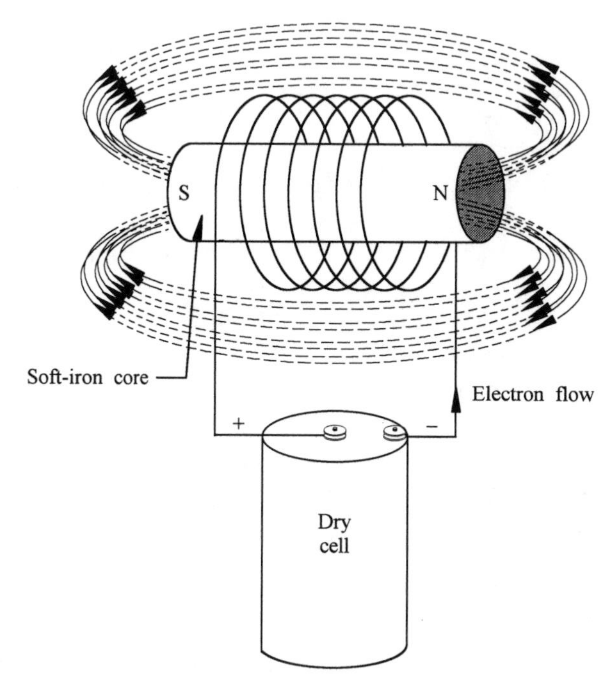

Figure 7-9. The soft-iron core greatly increases the strength of an electromagnetic.

coil produces a magnetic field that attracts the iron armature. The armature is attached to switch contacts—in this case, a normally closed and a normally open pair. You will find relay action applied in many control circuits in air conditioning and heating systems.

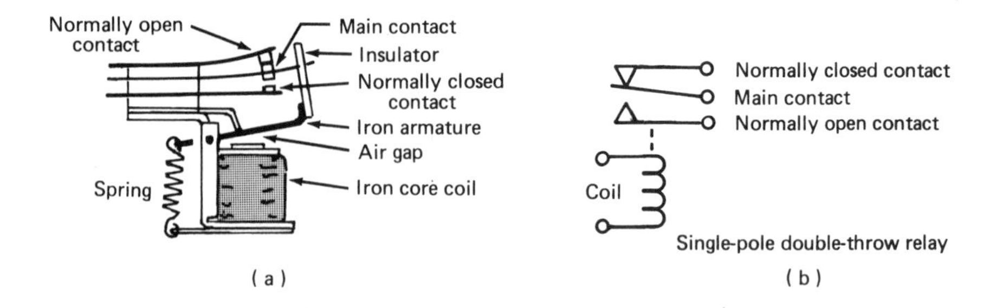

Figure 7-10. Construction of a relay.

SUMMARY

1. Magnets have a north and a south pole.
2. In magnets, like poles repel and unlike poles attract.
3. An iron rod can be magnetized by stroking a permanent magnet.
4. A current-carrying conductor produces a magnetic field around the conductor.
5. An electromagnet is produced when a current flows through a coil of wire.
6. An electromagnet attracts iron in the same fashion as a permanent magnet.
7. An iron core in an electromagnet increases the strength of the magnet.
8. An electromagnet will pull an iron core into its center, because lines of force try to shorten themselves and take the path of least resistance.

Chapter 8

Electromagnetic Induction

ELECTROMAGNETISM

We know that electricity produces magnetism in an electromagnet. Now we find that magnetism can produce electricity. Figure 8-1 shows how a voltage is *induced* in a coil when a permanent magnetic is moved in and out of the coil. This process is called *electromagnetic induction*. The are three important facts for us to notice about the demonstration:

- When the permanent magnet is not moving in the coil there is no voltage induced.
- When the magnet is moved out of the coil the voltmeter shows a voltage in one polarity (current in one direction).
- When the magnetic is moved into the coil the voltmeter shows a reading in the opposite polarity (current in the opposite direction).
- Voltage and current are induced as the magnetic field cuts the coils of the inductor. The direction of the current in the coil is determined by the left-hand rule for coils and generators (that was presented in Figure 7-10 of Chapter 7.

Another important form of electromagnetic induction is called the *self-induction* of the coil. We can observe this when a switch is opened in a circuit with a large current through a coil or motor winding. This self-induction is illustrated in Figure 8-2. We will find that a greater self-induction voltage is induced in a coil with an iron core than one with an air-core. This self-induction action is caused by the collapsing magnet field cutting the conductors of the coil. This action causes large voltage surges that produces arcing across switches, relays, and contactors, causing the contacts to burn and pit.

ALTERNATORS AND AC GENERATORS

A generator uses electromagnet induction to produce an output voltage. In a generator a magnetic field is moved across a coil or the coil is moved across the magnetic field. Generators, or alternators as they are called on automobiles, produce an alternating output in the form of a sine wave voltage. Figure 8-3 illustrates the action of a generator. A one-turn coil is seen in four positions of its rotation of 360 degrees.

(a) the conductor is moving in the same direction of the lines of force and no lines are cut.

(b) the conductor is moving across the lines of force and a positive voltage is induced (from the left-hand rule).

(c) the conductors are again moving in the same direction of the lines of force and no voltage is induced.

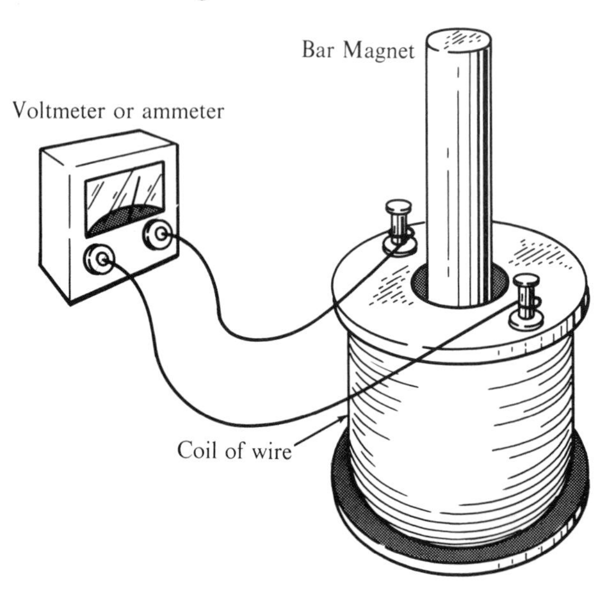

Figure 8-1. An example of an electromagnetic.

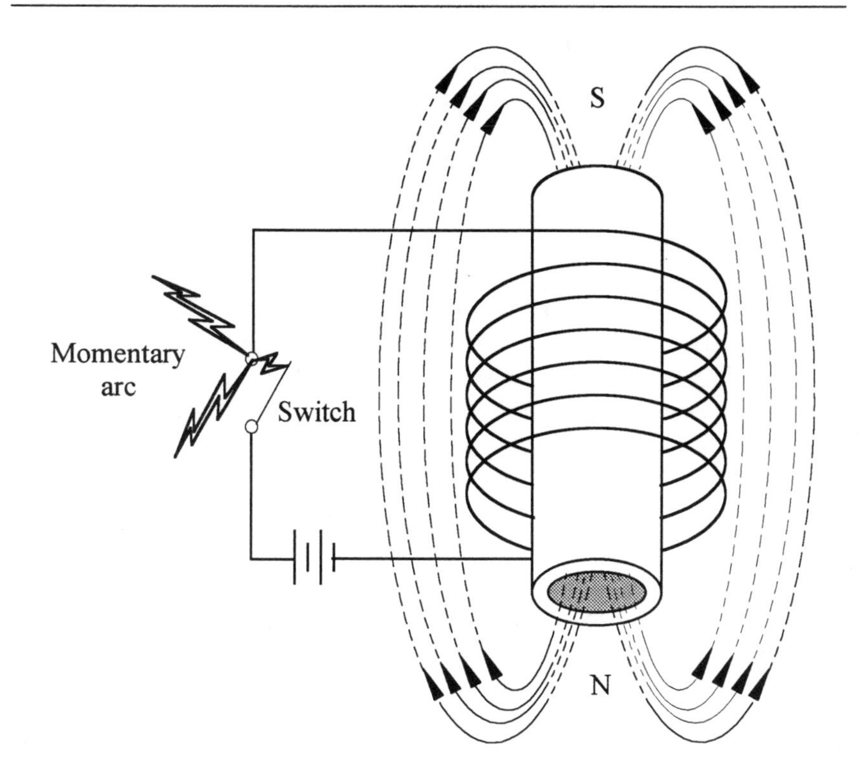

Figure 8-2. Self-induction causes arcing at the switch contacts.

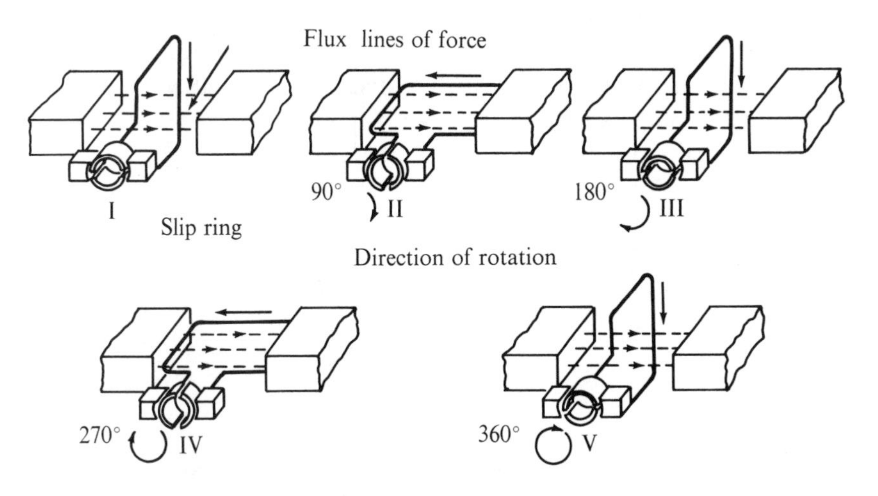

Figure 8-3. AC generator action.

(d) the induced voltage has once more risen to the maximum amount, but we notice that the polarity of the voltage has reversed.

(e) the conductors have returned to their starting point, and there is no voltage induced.

The coil has completed one revolution and produced a sine-wave voltage. The frequency of the voltage is determined by the speed of revolution of the coil, and is given in *cycles per second*. We specify cycles per second as Hertz (Hz). For example, the voltage used in our houses and laboratories is usually 120 volts at 60 Hz, and is produced by large generators driven by water power, atomic energy or gas.

TRANSFORMERS

A transformer can be compared with an ignition coil of an automobile except a transformer is operated on AC voltage and current. Transformers are used in all branches of electricity and electronics. The transformer shown in Figure 8-4(a) is a power transformer mounted on a power pole. Figure 8-4(b) is a 24-volt control transformer used to control the relay action of the gas valve in a heating unit.

An illustration of a transformer construction is shown in Figure 8-5. The transformer has an iron core with many turns of fine wire on the primary and few turns of heavy wire on the secondary. The size of the wire determines the current-carrying capacity of the windings. Transformers are rated for both current and voltage, and in volt-amperes.

The schematic diagrams of the transformers in Figure 8-6 represent (a) a step-down transformer, and (b) a step-up transformer.

In Figure 8-6(a) the primary has 100 turns and 117 volts, and the secondary has 1000 turns and 1170 volts. In Figure 8-6(b) the transformer is reversed with the primary having 1000 turns and 117 volts, and the secondary having 100 turns and 11.7 volts. The formula for the voltage and turns relationship is:

$$Np/Ns = Ep/Es$$

The power furnished by the primary supply is determined by the load on the secondary.

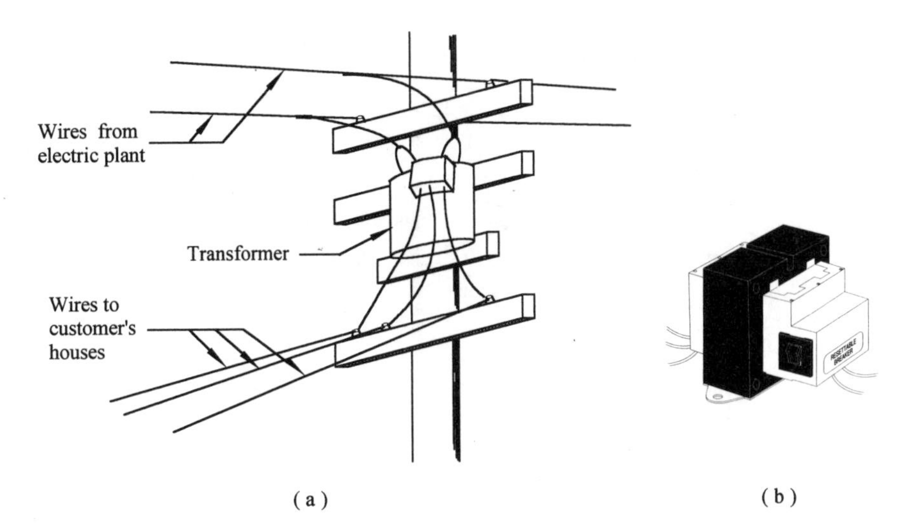

Figure 8-4. Examples of transformers: (a) a residential power transformer, (b) a 24-volt step-down transformer.

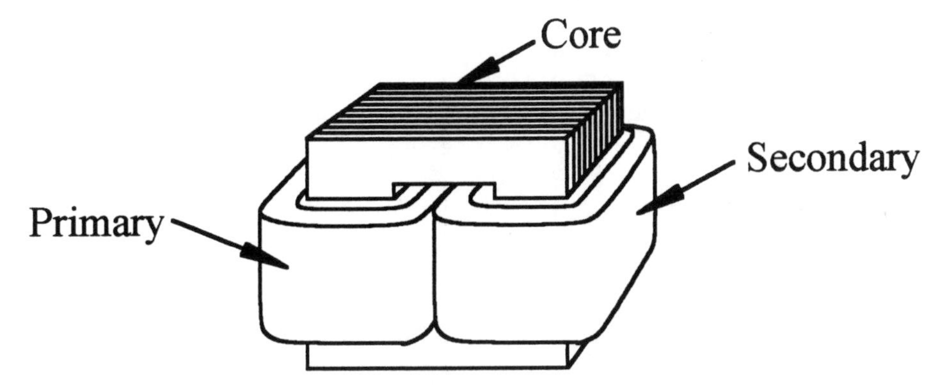

Figure 8-5. The construction of a transformer.

Example: A 24-volt furnace transformer has a load of a 10 ohm gas solenoid. The current and power in the secondary are:

- by Ohms law 24v/10 ohm = 2.4 A
- by the power law, P = I × E = 2.4A × 24 v = 57.6 W
- the power in the primary = power of the secondary
- the current in the primary is: I = P/E = 57.6 W/120 v = 0.48 A or 480 mA

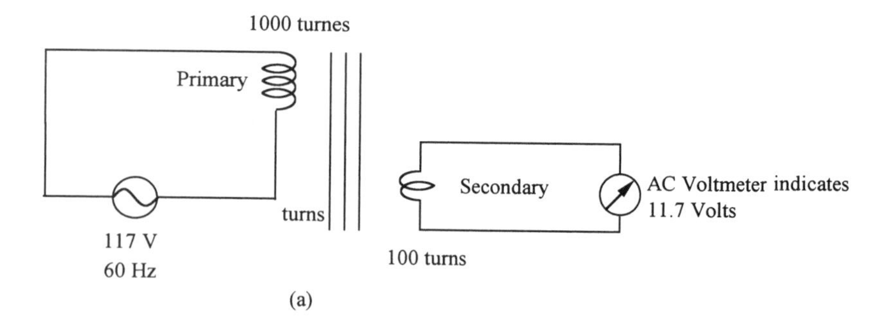

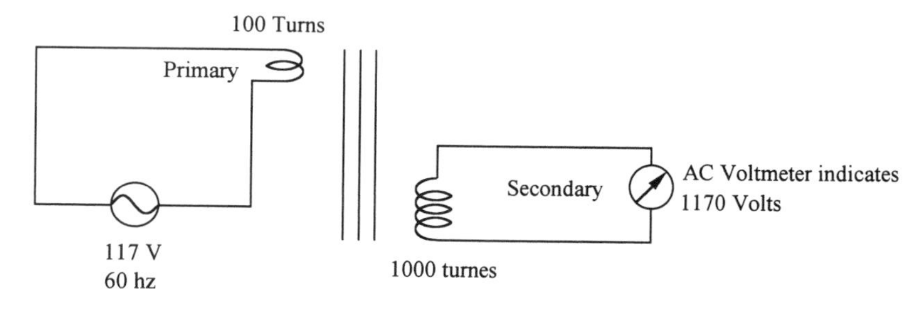

Figure 8-6. Schematic diagrams of (a) a step-down transformer, and (b) a step-up transformer.

POWER DISTRIBUTION

Electrical power is produced by generators that are turned by: water power, wind power, or gas and coal-fired turbines. The mechanical energy is converted into electrical energy to power residential and industrial applications. Figure 8-7 depicts and example of power distribution from the power plant to a residential application.

TROUBLESHOOTING COILS AND TRANSFORMERS

Both coils and transformers can be tested for open winding with an ohmmeter.

- To test a coil for open, remove the power and connections to one end of the coil, and place the ohmmeter across the coil. The meter should read a low resistance, that of the coil air.

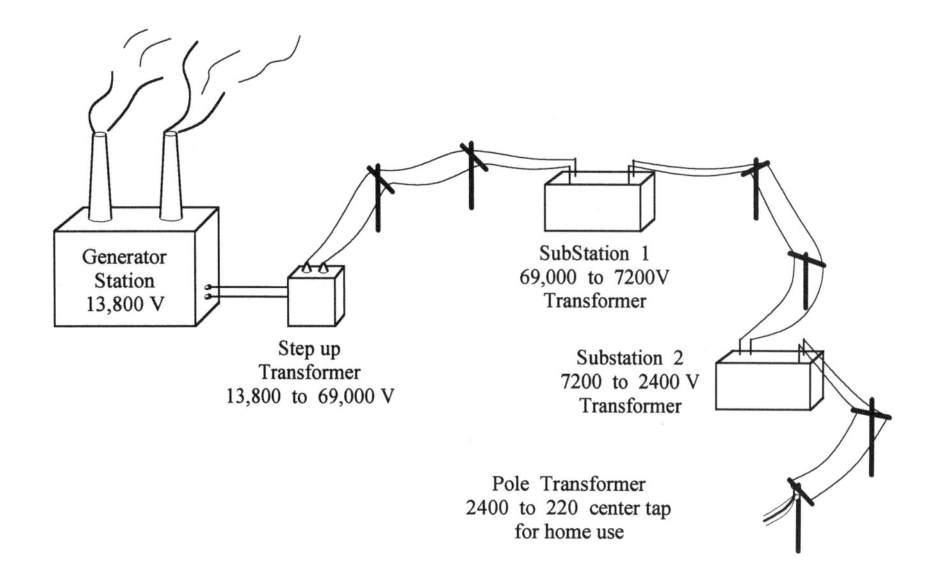

Figure 8-7. An example of a power distribution system.

- To test a transformer for open winding, turn the power off and disconnect the connections to one end of each of the windings. The ohmmeter should read a low value of resistance. A typical transformer will read approximately 5 to 10 ohms on the primary. The secondary reading will depend upon the winding current and voltage rating. A high current rating will read a low resistance and a high voltage rating will read a higher resistance.

- A short circuit on the secondary of a transformer will result in excessive primary current and a tripped circuit breaker or blown fuse. This type of problem can be isolated by removing the loads on each of the secondary windings one at a time. When the shorted load is removed the transformer will no longer trip the circuit breaker. You can then trace down the short circuit in the secondary circuit.

- An internal short between windings in a transformer will cause high line current and blowing of the fuse or the circuit breaker. There is no ohmmeter test for a shorted internal winding.

Whenever possible, a bad transformer or coil should be replaced with the same type. However, a transformer can be substituted if it has the same primary and secondary voltage rating and the current ratings are equal or higher than the original unit.

SUMMARY

1. A current and voltage is introduced in a wire or coil with a magnet when there is motion between the magnet and the conductors.
2. Anytime there is a current flow there is a magnetic field present.
3. Self-induction causes arcing across switch, relay, and contactor contacts.
4. Transformers operate on the principle of electromagnetic induction.
5. The power delivered to a load by the secondary of a transformer must be supplied by the primary supply.
6. Transformers can be classified as step-down, step-up or isolation transformers.
7. The power rating of a transformer is determined by the wire size of the transformer and its voltage rating.
8. Transformers can be tested with an ohmmeter and a megger.
9. When possible, a transformer or coil should be replaced with a factory replacement.

PROBLEM SOLVING

Draw the schematic of a 120-V to 24-V transformer.

- What would the output voltage be if you installed the transformer backward?
- What is the effect of an internal short between coils of the same winding in a transformer?

Chapter 9

Electrical Control Devices

INTRODUCTION

Refrigeration, air conditioning and heating systems use many mechanical and electro-mechanical devices. These devices are used for both operation and safety. Their functions are to:

- protect the technician and user,
- protect the circuits and components,
- protect the system,
- control the operation of the equipment.

The most important control devices are those that disconnect the electricity from a unit or system. Whenever possible, the technician must make repairs with the power disconnected from the system to prevent possible electrical shock. This is accomplished by the technician's personally disconnecting the main power switch or circuit breaker to the system.

The circuit breaker or switch should be opened and labeled with a red warning tag. The tag should be signed and dated. If the main has removable fuses they should be removed and placed in the technician's pocket. The author has had occasion to comply with all these suggestions and still was almost killed by an apprentice who removed the tags and replaced the fuses.

Before touching any deenergized unit, the technician should test for voltage with a voltmeter or other voltage indicator. For final safety the electrical circuit should be touched with the back of your hand. If the fuse of a circuit or system continues to blow there must be a circuit short or circuit overload.

This chapter deals with the operation and testing of these devices. You will learn how they operate, and how to test their operation. It is necessary to understand these devices, and to be able to follow their operation in electrical circuits and systems with circuit diagrams.

MANUALLY OPERATED SWITCHES

The purpose of a switch is to turn on or off the energy to an electrical circuit. The simplest type of switch is the single-pole-single-throw (SPST) switch shown in Figure 9-1. This switch has two positions and only one pole; it is either closed or open. This switch is usually enclosed in a plastic case and is called a *toggle switch*.

The toggle switch may be configured in a number of ways. An example is shown in the diagrams of the *single-pole-double-throw* (SPDT) configuration which can switch two circuit actions, but only one at a time. On the other hand, a *double-pole-single-throw* switch can be used to switch two circuits at once. A more complicated configuration is the double-pole-double-throw (DPDT) switch. It can switch two circuits at

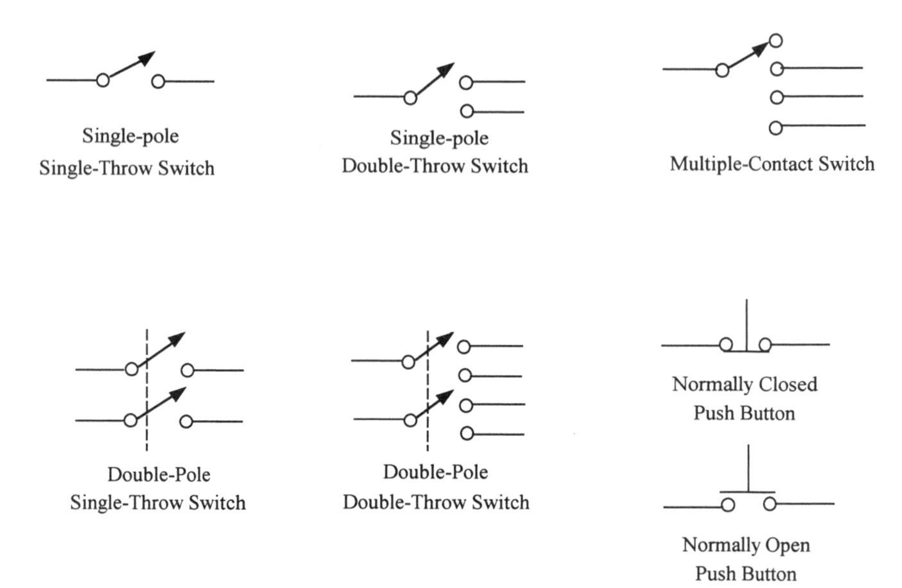

Figure 9-1. Schematic symbols of switches.

once either off or on. More complicated configurations such as a triple-pole-double-throw (TPDT) are possible. The toggle switch is sometimes configured for momentary action where the switch must be held in position. An example is shown in Figure 9-2(a). An example of the momentary push-button switch is the door switch for a refrigerator light. In cooling and heating systems, this type of switch is often used as an interlock safety switch where a panel must be closed to press the switch to complete the power circuit. Interlock safety switches must never be short circuited to make a unit operate. It is, however, sometimes necessary to activate an interlock switch by hand to troubleshoot a system.

When more complicated switching action is needed, than that achieved with the toggle switch, the wafer switch is used. This configuration can be used to switch several circuits. This switch may have as many as 20 positions and be stacked several wafers high on one shaft. This type of switch is rather difficult to troubleshoot; however, it seldom fails.

FUSES

In older equipment, fuses are the most common circuit protection device. Fuses are selected to protect the wiring and equipment from excessive current. Actually, a fuse is a low wattage, low value resistor. When excessive current flows through the resistance of a fuse the power

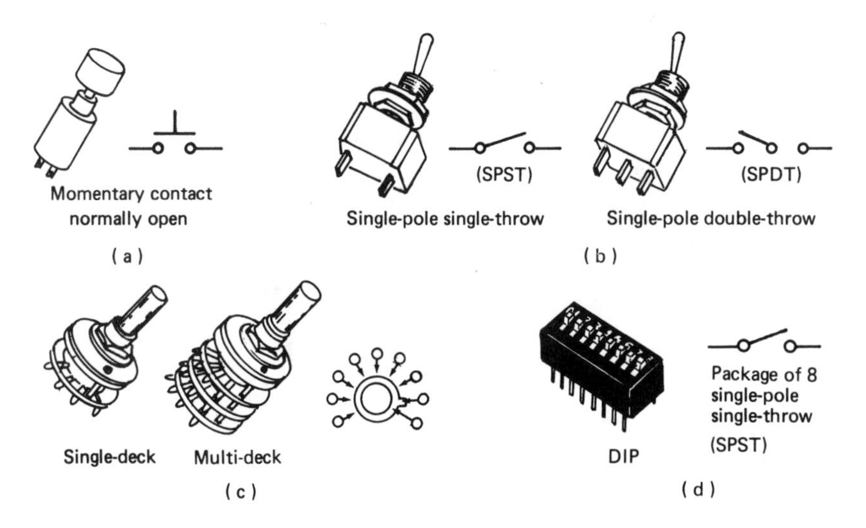

Figure 9-2. Examples of switch packaging.

produced causes it to "burn open." Fuses are manufactured in a number of shapes and sizes. Each fuse has specific current and voltage ratings. Some fuses, called *slow-blow,* are designed to allow short periods of overcurrent without opening.

Figure 9-3 depicts several different types of fuses. Some larger cartage type fuses have replaceable internal links.

Some safety features for replacing fuses are:

1. Fuses should always be removed and replaced with an insulated fuse puller, with the power switch in the off position.

2. A fuse should always be replaced with the same type having the same current and voltage ratings.

3. A fuse must never be replace with one with a higher current rating.

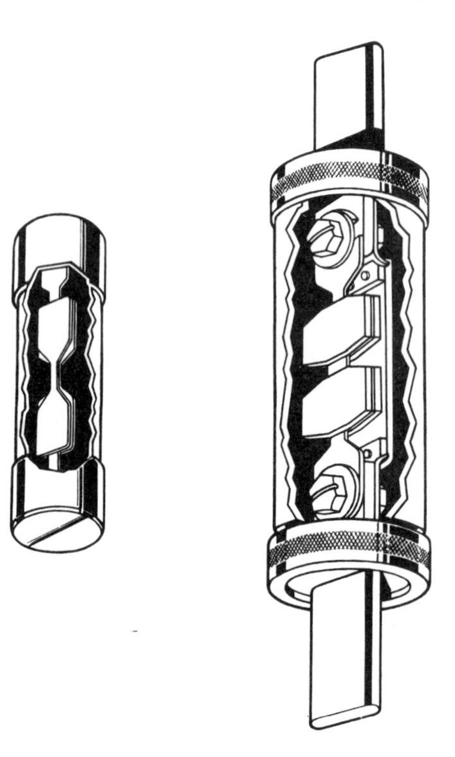

Figure 9-3. Example of fuses.

CIRCUIT BREAKERS

Circuit breakers are electrical-mechanical devices that automatically disconnect the power in the event of an overload. There are two basic types of circuit breakers: the heat-overload beaker and the magnetic overload breaker. The heat-overload breaker is found in older systems, while the magnetic release breaker is more common in newer installations. A typical thermal-circuit breaker is shown in Figure 9-4.

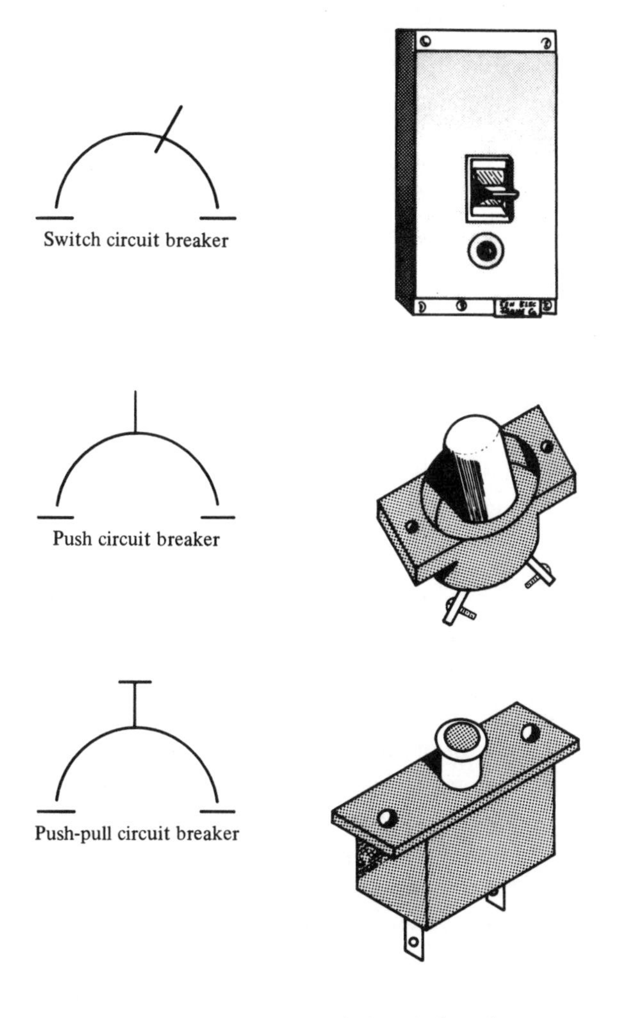

Figure 9-4. Examples of circuit breakers.

Magnetic-circuit breakers operate on the principle of a solenoid. Line current flows through the control coil. Excessive line current "trips" the operating arm opening the contacts.

Heat-overload circuit breakers use an internal bi-metal strip as a disconnect. When the current rating is exceeded the bimetal strip heats. One of the metals expands greater than the other and forces the contact to bend, opening the contacts and tripping the breaker. Once a circuit breaker has tripped, the handle must be pushed into the reset position to again complete the circuit. Both magnetic and bi-metal heat overload circuit breakers are manufactured for 110 volt, 220 volt, and 440 volt operation, and in many current ratings.

Circuit breakers may fail due to continued use or age. The safe current capacity of an electrical unit and/or the wire size determines the maximum current rating of a circuit breaker. For example, a circuit connected with number 12 copper wire must never be controlled with a breaker of greater than 20 amperes. Circuit breakers should never be used as switches.

RELAYS

Switches fail due to frequent operation, or because they are required to switch on and off high levels of current. Each time a high current is switched on and off the inductance in the lines and circuits cause a very high voltage arc across the contacts. This arc causes pitting of the contacts and finally failure of the switch to make continuity in the circuit. The partial answer to this problem is the relay and relay circuit shown in Figure 9-5. The relay is an electrical switch. A small current through the winding of the coil causes a magnet field that attracts the armature to the coil. The moveable *armature* and the fixed *stator* have high current contacts that touch to complete the circuit. These high-current contacts are usually made of silver, and are designed to operate many more times and at much higher currents than switches.

Another advantage of relays is that we can switch a high current or voltage a distance from the low current and low-voltage switch that controls them. Relays may have contacts designed as double-pole-double-throw (DPDT), single-pole-single-throw (SPST), normally open (N.O.), and normally closed (N.C.). The N.O. and N.C. designations define the contact condition when the relay is not energized.

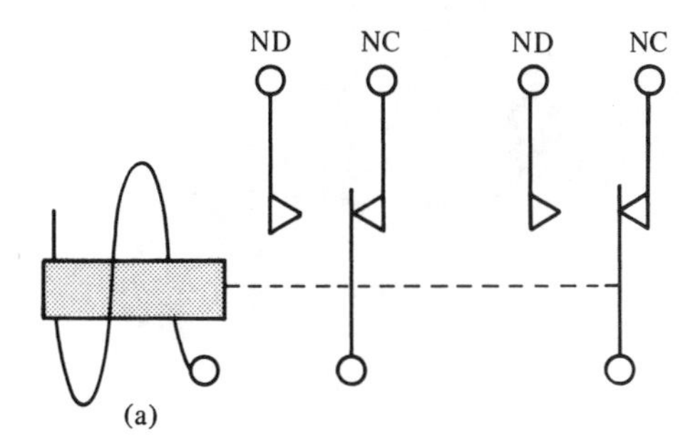

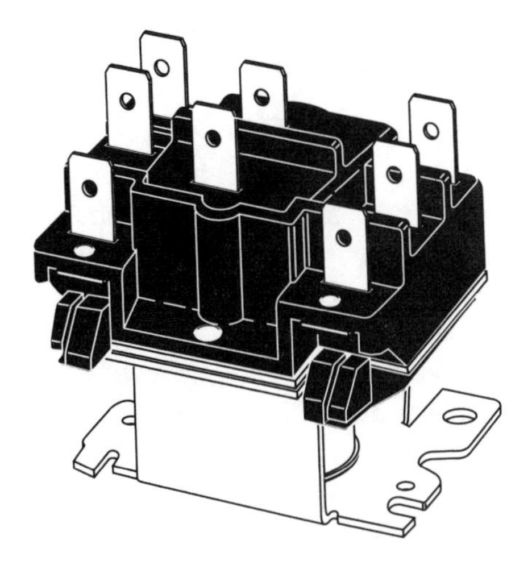

Figure 9-5. Example of a relay: (a) diagram, (b) configuration.

CONTACTORS

Contactors are similar to relays; however, they are designed to handle higher currents. In air conditioning and refrigeration circuitry systems contactors are used to switch the main loads, such as compressor motors. Another distinction of contactors is that their coils are replaceable, whereas relay coils are fixed. Figure 9-6 depicts an example of a high current contactor. An input of 120V at 1 A operates the contactor that can control a three-phase 440-V motor that draws a current of 40 A.

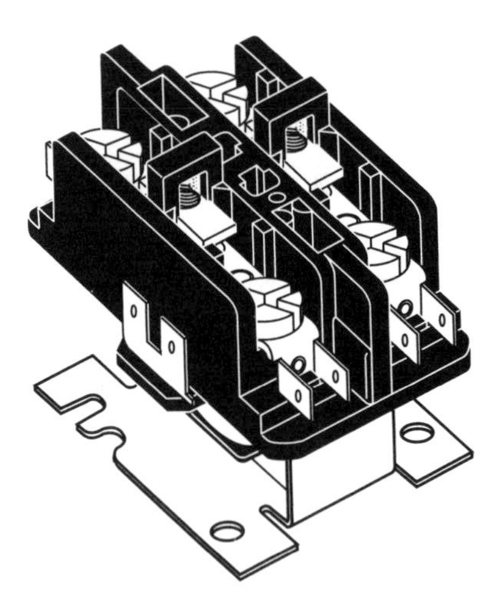

Figure 9-6. A high-current contactor device.

SOLENOIDS

Solenoids are electrical coils that use electromagnet action to pull a movable core into the coil. The movable core is attached to a lever that controls a mechanical action. Figure 9-7 shows how a solenoid's mechanical action can be used to open or close vents or vanes in air ducts. When a voltage is applied to the coil the magnetic action of the coil pulls the plunger into the coil, and the coupling mechanism moves the arm and activates a mechanism.

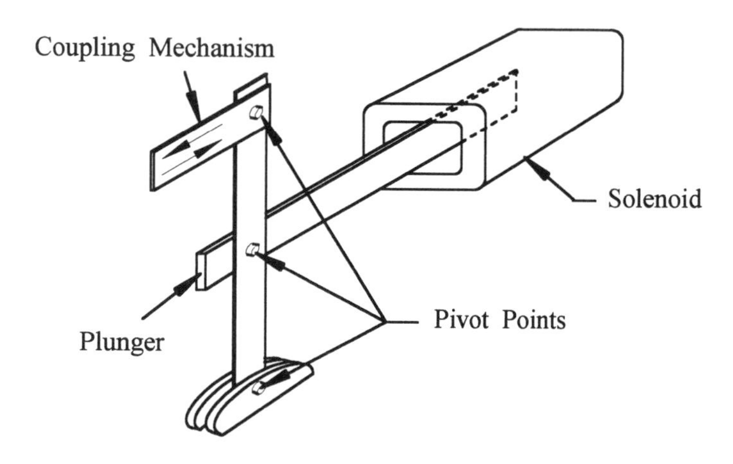

Figure 9-7. Example of a solenoid action.

MOTOR OVERLOAD PROTECTION DEVICES

Compressor motors are designed with internal or external overload protection devices. A compressor motor may become overloaded by an event that causes excessive demand and results in excessive current. Excessive demand can be caused by over cycling, a motor too small for the job, incorrect line voltage, or improper ventilation or a short circuit. All of these overloads cause excessive current and heat. Overload protection devices are designed to sense one or more of these conditions and disconnect the motor to prevent burn out. Motor protection of a compressor motor is designed to operate at 70 to 80 percent of maximum current and lower than fuse or circuit breaker ratings.

Fuses and circuit breakers typically have a rating of 150 to 160 percent of rated load current. However, PSC type compressor motors can be fused at higher levels.

Bi-metal type motor protectors are often used to limit current for single-phase hermetic type compressors. These heat-sensitive overload protection devices are placed to instantly detect motor temperature and disconnect power. If the compressor overheats, the bimetal strip in the overload device bends and opens. The compressor must cool before the overload device resets itself. In some cases the overload device must be reset by hand before the compressor motor is returned to service. Figure 9-8(a) is an example of the overload device in a compressor circuit.

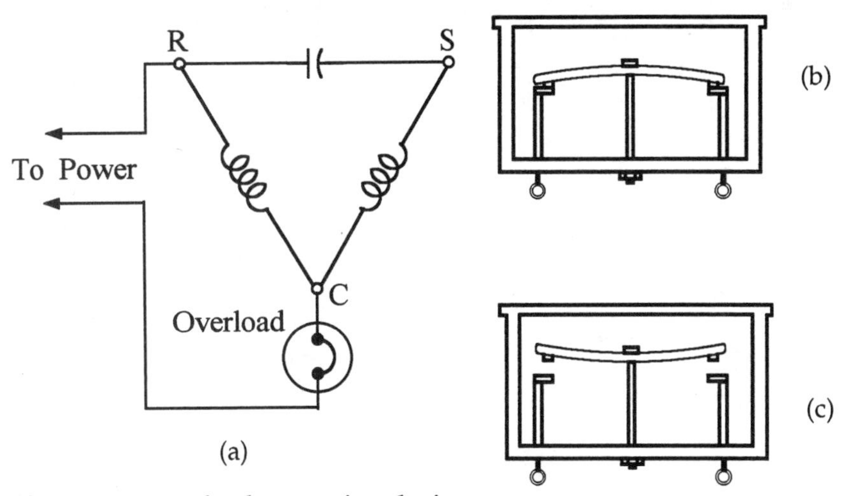

Figure 9-8. Overload protection device: (a) circuit application, (b) normal position, (c) heated position.

Some heat-overload devices depend only upon compressor temperature to heat and open. Others operate in the same manner but have an internal heater for a quicker response. Both these devices are used on fractional horsepower motors. Figure 9-9(a) shows a compressor circuit with a device that protects the motor from both excessive current and heat. If the motor gets too hot from excessive current, over cycling, or insufficient cooling, the overload device in Figure 9-9(b) is inserted in the starting winding and protects the motor from excessive temperature and excessive current during the starting cycle. Figure 9-8(a) is an example of the overload device in a compressor circuit.

The overload protection in many hermetic compressors is mounted inside the compressor shell and is shielded from the refrigerant. These types of overload devices are not replaced in field servicing. The faulty unit is usually replaced and the repair referred to a repair facility. On lower horsepower hermetic compressor motors the overload protection device is a bi-metal disc connected to the start terminal and located outside the compressor case.

Figure 9-9 shows an example of three-phase motor protection. This system uses a thermal overload relay for motor protection. Excessive current in any of the three legs of the motor winding results in excessive

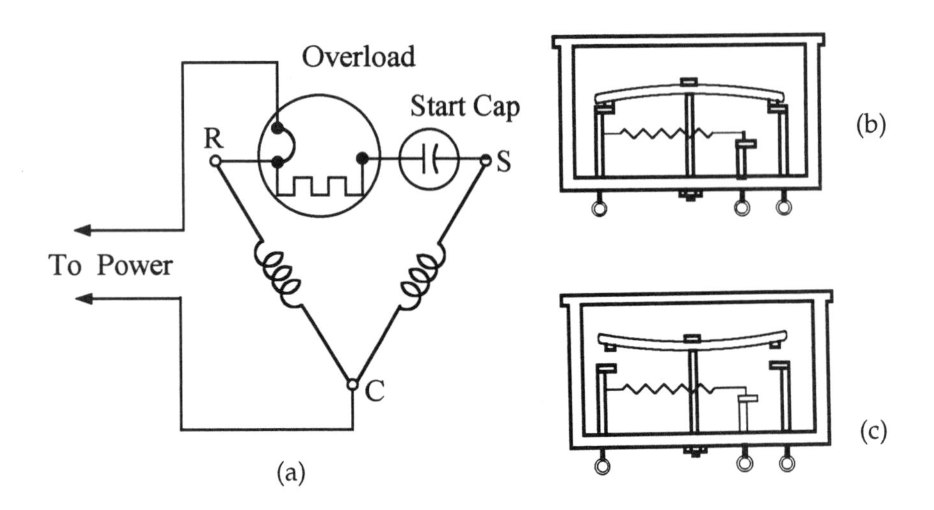

Figure 9-9. A compressor using both current and heat protection: (a) circuit diagram, (b) overload device in normal position, (c) in heated position.

heat and opening of the overload contacts. The three N.C. relay contacts are in series so that either will disconnect the 240-V coil of the control contractor. This removes power from all three windings of the motor.

Electronic motor protection modules offer the fastest and most complete motor protection. With this type system, heat sensors are placed at several positions in the motor. The sensors are connected to an electronic module that controls motor current, usually by controlling the contactor coil current.

Electronic control devices have the advantage over temperature control devices of being able to detect a number of abnormal conditions in the motor. For example, electronic control devices can detect out-of-phase current, low-winding current, high-winding current, over cycling, loss-of-phase current, and phase reversal. Figure 9-11 depicts an electronic motor control module. Repair of a faulty electronic control mod-

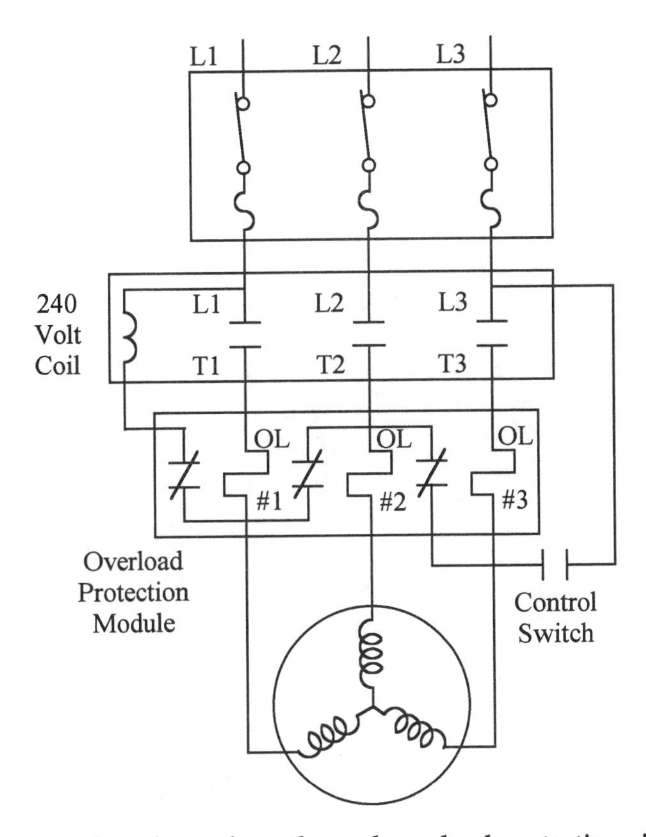

Figure 9-10. A three-phase thermal overload protection circuit.

ule is usually beyond the skills of the service technician. Such units must usually be replaced and the faulty unit repaired by a service center or returned to the manufacturer for repair.

COMMERCIAL AIR CONDITIONING CONTROLS

Commercial air conditioning systems have many controls to maintain proper operation and long life of the components. Figure 9-12 shows a diagram of a refrigeration system. The arrows in the system indicate the flow of refrigerant gas and liquid. The basic system is the compressor, condenser, and evaporator.

The controls from top to right are:

- **Temperature control**—Senses the temperature.
- **Head pressure control**—Controls the maximum compressor pressure, and helps prevent evaporator freeze-up, liquid slugged compressors.

Figure 9-11. An example of an electronic control module.

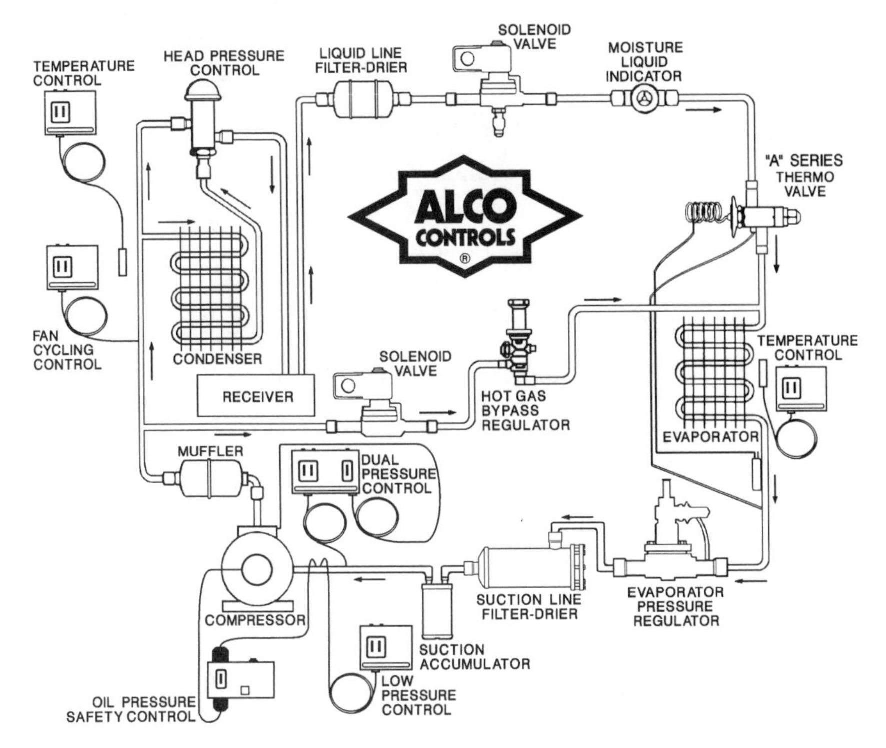

Figure 9-12. Example of controls in commercial refrigeration system.

- **Liquid line filter-drier**—Filters the refrigerant and removes moisture from the system (non-electrical).
- **Solenoid valve**—Electrical solenoid control valve.
- **Moisture liquid indicator**—Indicates the presence of moisture in the system through a view glass (non-electrical).
- **Fan cycling control**—Controls the cycling of the condenser cooling fan.
- **Thermostatic valve**—Controls the flow of refrigerant to the evaporator.
- **Temperature control**—Controls the temperature by engaging the compressor.

- **Muffler**—Reduces compressor noise (non-electrical).
- **Dual pressure control**—Limits both excessive high and low pressure and safeguards the compressor motor.
- **Evaporator pressure regulator**—Regulates the suction pressure of the gas back to the compressor (non-electrical).
- **Oil pressure safety control**—Limits excessive oil pressure.
- **Low pressure control**—Limits low oil pressure.
- **Suction accumulator**—Accumulates moisture from the suction line dryer (non-electrical).
- **Suction line filter-drier**—Removes moisture from the refrigerant in the suction line (non-electrical).

TESTING CONTROL DEVICES

Switches

The action of a switch can be tested by placing an ohmmeter across the two terminals and opening and closing the switch.

- The ohmmeter should read infinity for open and zero for closed. Any other reading indicates that the switch is bad an should be replaced. Testing with the ohmmeter must be accomplished with the power off. The wires to one end of the switch will often have to be removed to prevent a sneak circuit that might cause a low reading on the open position.

Fuses

Fuses can be tested by making a continuity test with an ohmmeter or, in some cases, a voltmeter.

- With the leads of the ohmmeter placed across a fuse the reading should be near zero. A fuse may be tested in or out of the circuit. *Power must be off for this test.*
- A fuse may be tested with a voltmeter with the power on and the load connected. One lead of the voltmeter is placed on the line side

of the fuse and the other on the load side of the fuse. A reading of zero volts indicates a good fuse link. A reading of line voltage indicates an open fuse link. These readings are inconclusive unless the load is connected and on.

Circuit Breakers

Circuit breakers may be tested by using an ohmmeter or a voltmeter.

- An ohmmeter reading, of a circuit breaker, should be zero in the closed or set position, and infinity in the off position.

- A circuit breaker may be tested with a voltmeter by placing the meter leads on the input and the out terminals. A good breaker will read zero volts since it will be a short circuit. An open will indicate line voltage. These tests are valid only if there is a load on the line.

- To test with a voltmeter the voltage can be measured between two of the lines. The input voltage to the circuit breaker should be measured first to assure that there is line voltage. Then one lead of the voltmeter is placed on one of the output terminals and the other on one of the other lines.

- For 120-V operation one of the voltmeter's test leads is connected to ground or neutral and the other to the output terminal of the circuit breaker.

- To test a 240-V or a 440-V breaker's measurements must be made at the output of each terminal. Voltage Measurements are made from one of the outputs of the breaker to one of the other input terminals of the breaker. Readings should be 120 volts for the 240-V system and 220-V for the 440-volt system.

- Figure 9-13 illustrates a method for testing fuses and circuit breakers on a live circuit.

Relays, Contactors, and Solenoids

Relays, contactors, and solenoids are tested for contact action with an ohmmeter much as if they were switches. Contact action can often be

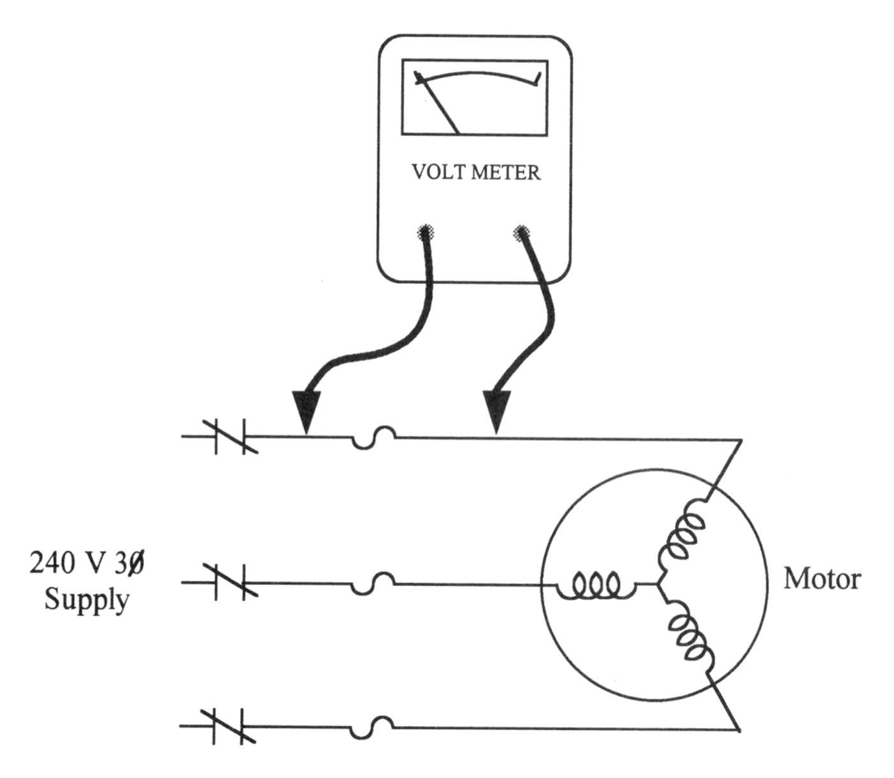

Figure 9-13. Tests of fuses or circuit breakers on a live circuit.

observed visually when one of these devices is energized. The power must be removed to test these devices with an ohmmeter. Figure 9-14 depicts the method of using the ohmmeter to test a relay.

- The coil of an electromagnetic device can be tested for continuity and resistance. A reading of infinity indicates an open coil. Resistance values are seldom indicated, on schematics, for these devices. The winding resistance of a similar device might be measured for comparison.

- With the power removed the armature can sometimes be controlled by hand, and the resistance across the contacts can be measured by connecting an ohmmeter across the contact terminals. The case may have to be removed to perform this test. BE SURE THAT POWER IS OFF before attempting this test. The closed contact resistance should read zero ohms.

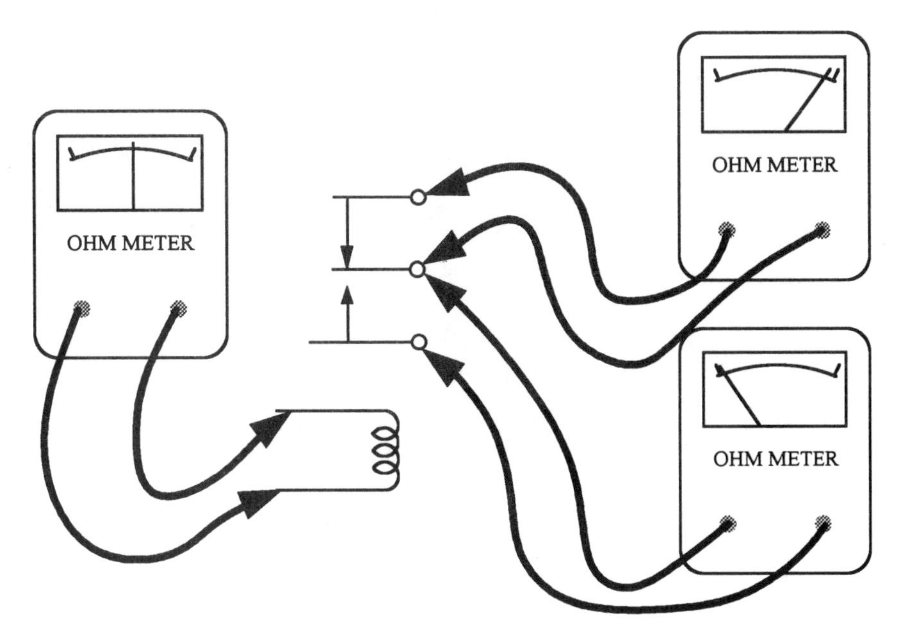

Figure 9-14. Testing a relay with an ohmmeter: (a) testing the coil, (b) testing the contacts.

- The coil and armature action can be tested by observing the throw action of the armature in an energized state. The case may have to be removed to observe the contacts.

- Relay, contactor, and solenoid contacts, that have been burned and pitted, can sometimes be cleaned with a special file called a *burnishing tool*. This device has a very fine abrasive surface that will not leave ridges on the contact surfaces. Very fine emery cloth or number 000 sandpaper can also be used.

- Whenever possible, replace a relay, contactor, or solenoid with the manufacture's suggested replacement. However, if an exact replacement is not available you may make a substitute under the following conditions.

- The switching coil must utilize the same voltage and current as the original.

- The contacts must have an equal or higher current rating than the original relay.
- The contacts must be of the original configuration, for example, DPDT, N.O., N.C., or TPDT.

SUMMARY

1. When working on electrical equipment the technician should red tag the fuse or circuit breaker box.
2. Switches are the primary safety device to prevent electrical shock.
3. Fuses and circuit breakers are overload safety devices.
4. Fuses and circuit breakers must never be replaced with ones with a larger current rating.
5. Solenoids, relays and contactors are electrical switches.
6. The contacts of a relay, solenoid or contactor can be tested with an ohmmeter.
7. The contacts of a relay, contactor or solenoid can be tested with a voltmeter

QUESTIONS

1. Explain how you would test the coil and the contacts of a relay or contactor.
2. What is the difference between a coil and a contactor?
3. You are about to remove a relay from an air conditioner; what precaution should you take?
4. You are working on a defective air conditioner that requires line voltage to make voltage measurements; what precautions should you take?
5. You must replace a relay or contactor, but the exact unit is not available. How would you select a replacement?

Chapter 10

Semiconductor Devices

INTRODUCTION

In this chapter you will learn the theory of semiconductor materials and how they function as rectifiers and control devices. Each of the devices introduced here can be found in the environmental control system. It is therefore necessary that you have a working knowledge of their function, and the correct method of testing for faulty devices. The operation of each device will be presented in a typical circuit. A summary of testing of all the devices is presented at the end of the chapter.

A semiconductor is a material that is neither a good conductor nor a good insulator. We can compare a semiconductor with a resistor, although we will find that it is a special kind of resistor. The most important use of semiconductor materials is in *control devices*. Control devices are components that are used to control the direction of current flow or to increase the strength of current or voltage.

The two most often used semiconductor substances are *germanium* and *silicon*. The materials germanium and silicon are crystals. This means that their atoms are lined up exactly in rows and columns like a battalion of soldiers. In the beginning of semiconductor manufacturing, germanium was the most often used. However, silicon is currently universally used because of its high temperature rating.

Both germanium and silicon have four electrons in their outer orbit. A simplified picture of a silicon atom with its four outer electrons is shown in Figure 10-1. When silicon is melted and allowed to cool, it forms into a solid in which each atom shares (attracts) four silicon atoms in the crystal as shown in Figure 10-2. That is, the outer shell of each atom is filled with eight electrons, although four of the electrons are not its exclusive property. This electron sharing is said to be caused by *covalent forces*. These covalent forces cause a lineup of electrons from nearby silicon atoms.

A careful study of Figure 10-3 shows that the positive electrode attracts an electron from the left-hand silicon atom. Now this atom is left with a missing electron. The location of this missing electron is called a *hole*. We see that if an electron jumps into the hole from the next atom to the right, this jump will leave a hole behind. Therefore, we have hole travel in one direction in a silicon crystal, as well as electron travel in the other direction. The result of electron and hole travel is called *drift current*. In the case of a pure crystal, we call this saturation current, because there are very few electrons produced by the surrounding light and heat. The available current cannot be increased by an increase in the supply voltage. The only way to increase the available current is to apply more light or heat to the crystal.

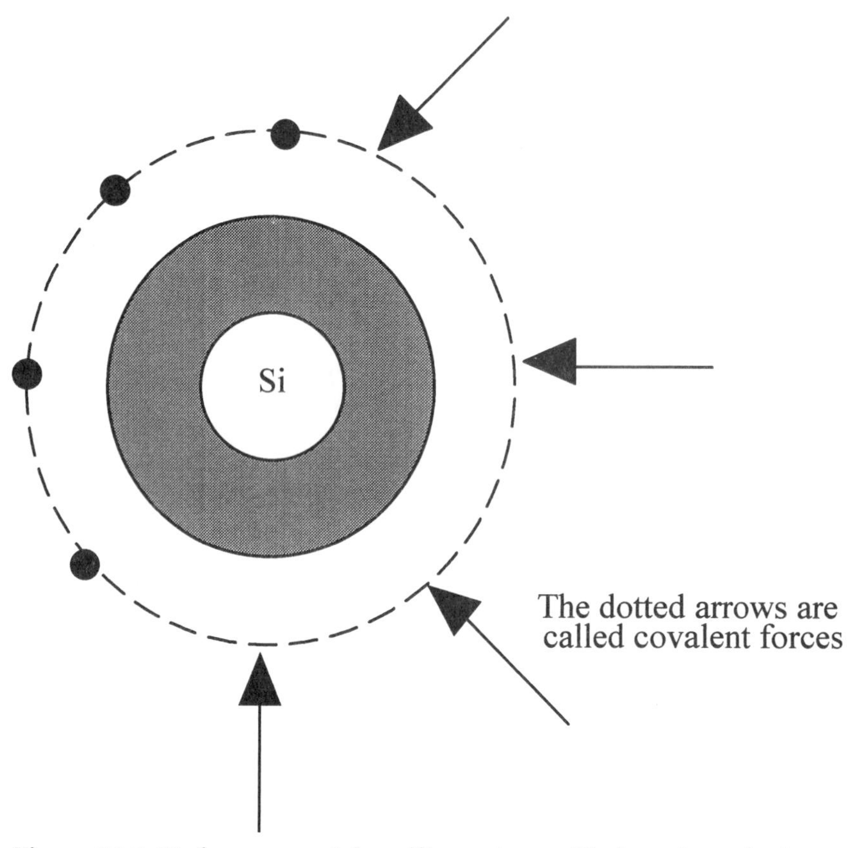

Figure 10-1. To form a crystal, a silicon atom will share four electrons from nearby atoms.

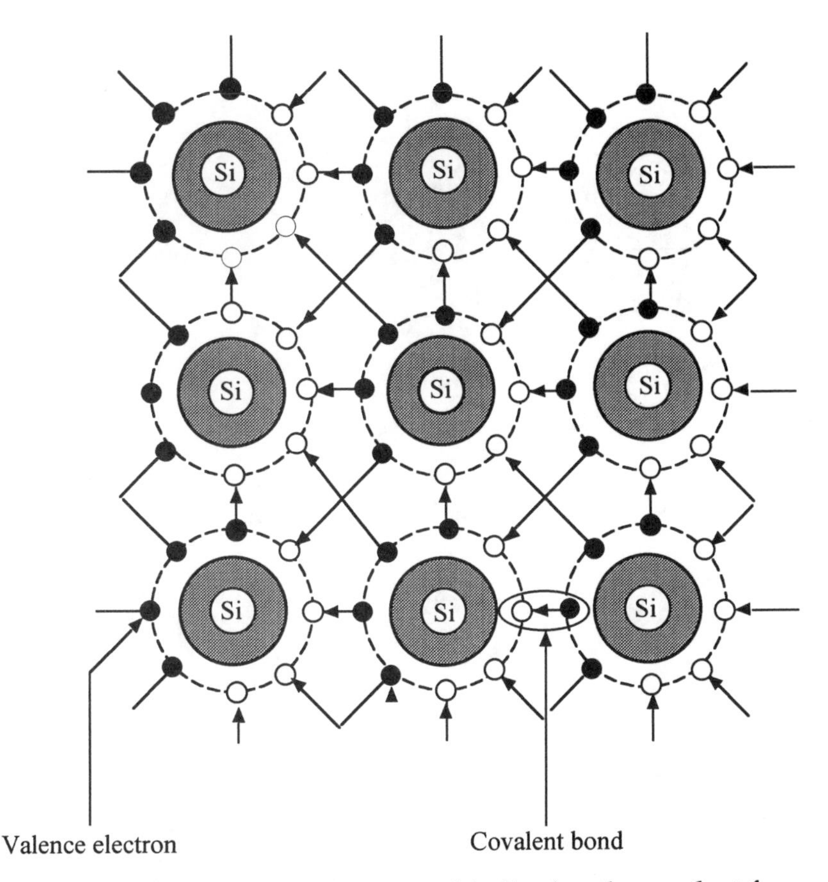

Figure 10-2. Structure of a silicon crystal indicating the covalent forces between atoms.

Example 10.1

Connect a silicon diode in series with a microammeter and a 3-volt battery as shown in Figure 10-4. It is important that the diode be connected in the circuit with reverse polarity. In this condition, only a small reverse current flows, which is chiefly saturation current. When the diode is heated between your fingers or by a lamp or soldering iron, the saturation current increases. Overheating the diode will destroy it as a control device.

Figure 10-3. Saturation-current flow in a pure silicon crystal.

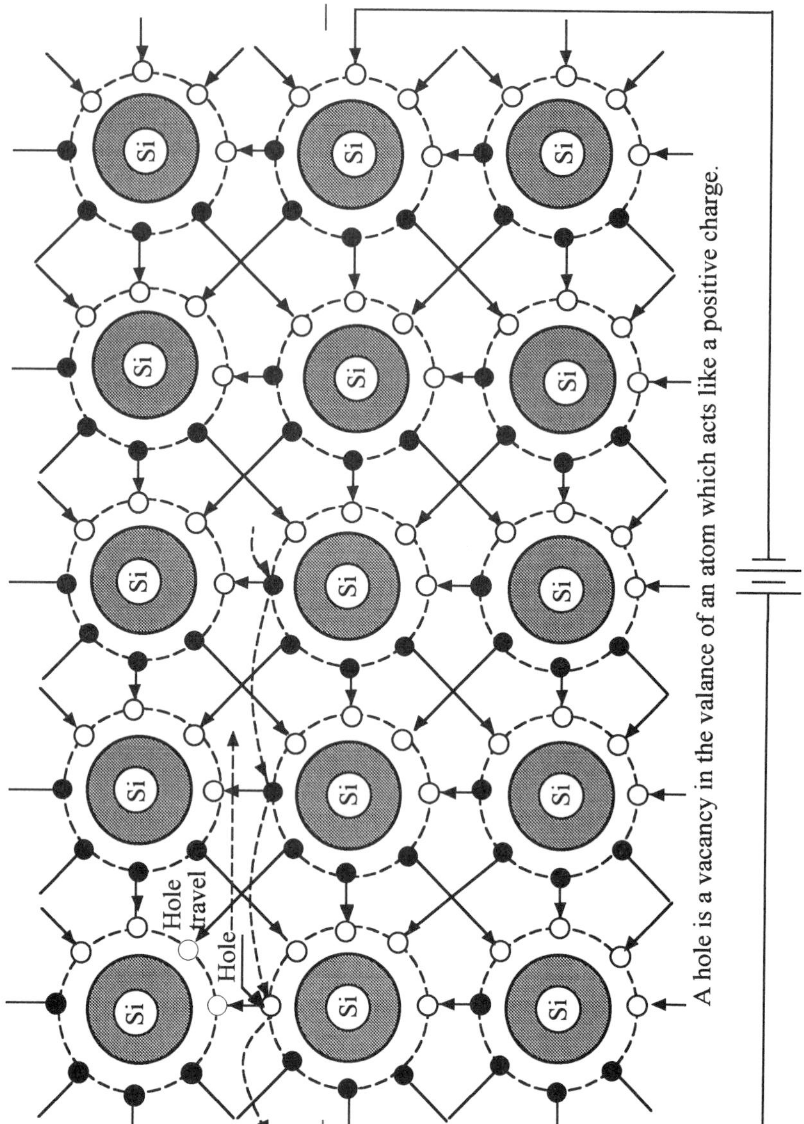

A hole is a vacancy in the valance of an atom which acts like a positive charge.

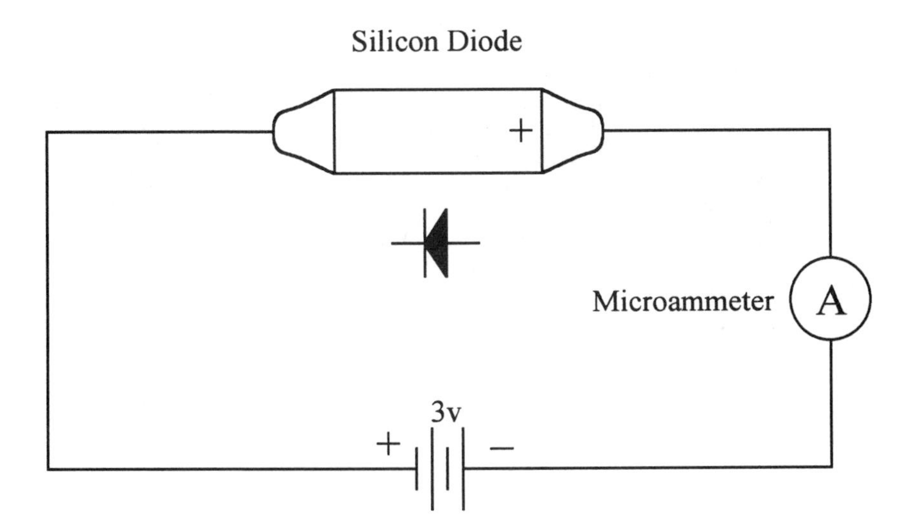

Figure 10-4. Saturation-current demonstration.

To develop a semiconductor material so that it operates as a control device, pure silicon or germanium are *doped*. The semiconductor is doped with a trace of phosphorus, a trace of boron, or various other "impurities." An impurity is any substance other than the semiconductor. Semiconductor devices will not operate properly without the addition of impurity atoms. Two widely used doping substances are boron and phosphorous.

We note from the plan of the boron atom, in Figure 10-5, that boron has three outer electrons or one less outer electron than silicon. On the other hand, phosphorus has five outer electrons or one more outer electron than silicon. If a trace of boron is added to a pure silicon material, the crystal that is formed will have a rather lower resistance than a pure silicon crystal. Also, if a trace of phosphorus is added to a silicon crystal, the new crystal will have relatively less resistance compared to the silicon crystal. Materials with three valence electrons are called *trivalent atoms*; and materials with five valence electronics are called *pentavalent atoms*. Be aware that silicon doped with trivalent atoms is quite different from silicon doped with pentavalent atoms. Let us see why this is so.

Again observe in Figure 10-6 that each silicon atom fits perfectly into the crystal arrangement. Each atom fits perfectly because each atom has four valence electrons. However, if a phosphorus atom is doped into

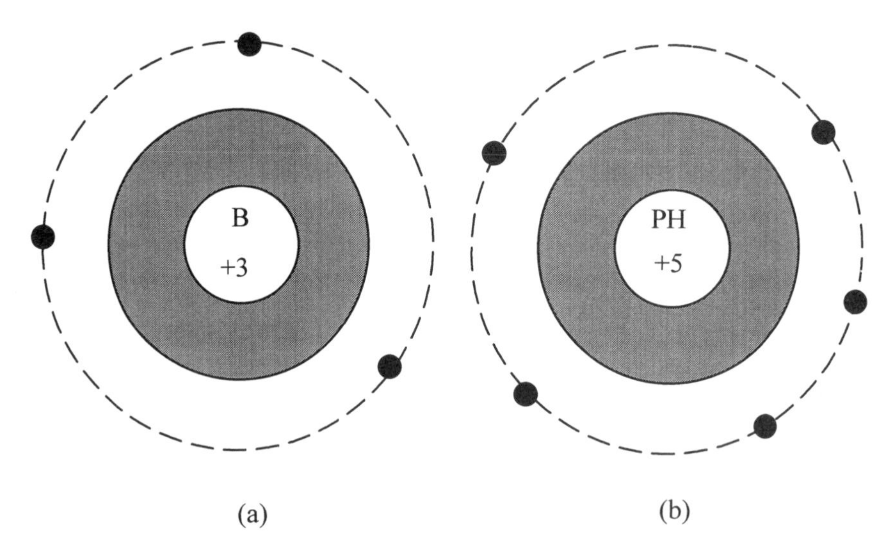

Figure 10-5. Semiconductor doping substances: (a) the boron atom has three outer electrons, (b) the phosphorous atom has five outer electrons.

the place of one of the silicon atoms, as in Figure 10-6, the fit is not perfect. This is because the phosphorus atom has an extra outer electron which does not fit into the crystal arrangement. Therefore the extra electron cannot be attracted by the covalent forces. Instead this extra electron floats through the crystal as a *free electron.* This free electron can move easily from one end of the crystal to the other end if a voltage is applied across the crystal. This type of semiconductor is called *N-type silicon.* N-type means that the semiconductor has free negative charges, or free electrons as current carriers.

When a voltage is applied across a piece of N-type silicon, electrons flow much the same way as marbles pushed into a pipe as shown in Figure 10-7. As with pure silicon, the amount of current will increase if we increase the temperature of the material because of the added saturation current. Now let us replace a silicon atom in the crystal with a boron atom, as shown in Figure 10-8. Notice again that the fit is not perfect, because the boron atom has only three outer electrons. This missing electron, or hole, is like a positive charge floating in the crystal. This hole can easily move from one end of the crystal to the other end when a voltage is applied across the crystal. This type of semiconductor is called *P-type semiconductor.* P-type means that the crystal has free positive

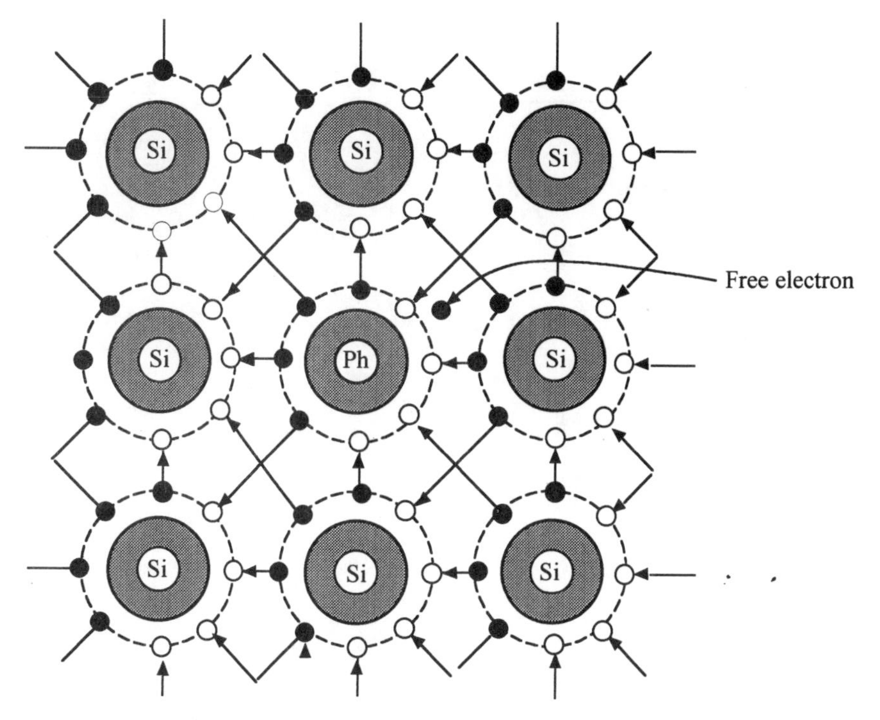

Figure 10-6. Silicon doped with phosphorus, allowing a free electron.

charges, or holes, as current carriers. When a voltage is connected across a p-type crystal, holes flow in much the same way as marbles flowing out of the pipe as was shown in Figure 10-7.

THE JUNCTION DIODE

A *junction diode,* or *rectifier* consists of a layer of N-type semiconductor and a layer of P-type semiconductor, as shown in Figure 10-9. That is, the N-type crystal and the P-type crystal meet at a junction. In the manufacturing of a semiconductor diode, the junction is continuous and there is no space or break at the junction. The junction is simply a place where the N-type material turns into P-type material. The atomic arrangement in a junction is shown in Figure 10-10. Notice that there is a phosphorus atom on the left-hand side of the junction, and that there is a boron atom on the right-hand side of the junction.

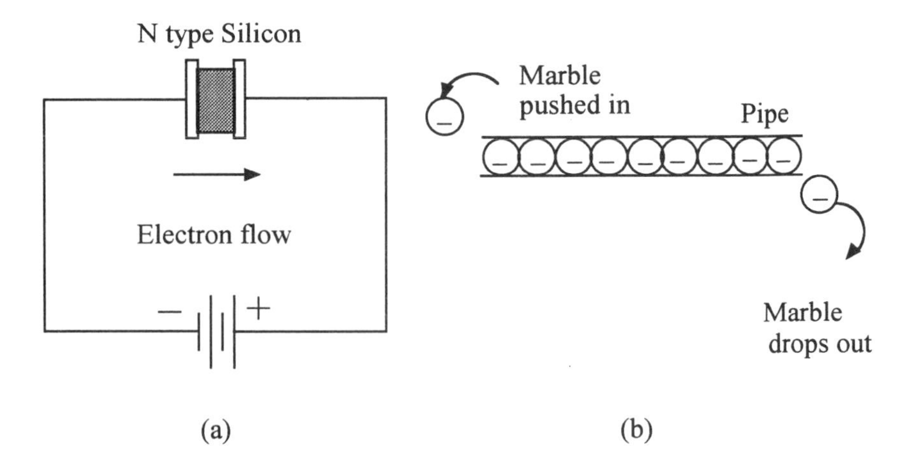

Figure 10-7. Marble-and-pipe comparison: a. basic circuit, b. marble-and-pipe comparison.

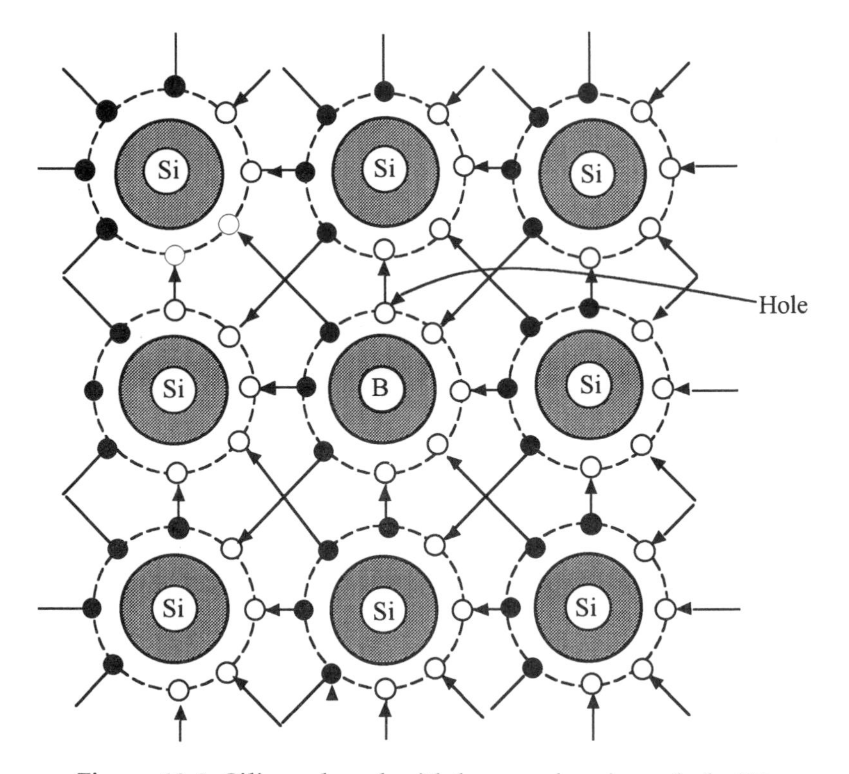

Figure 10-8. Silicon doped with boron, showing a hole (H).

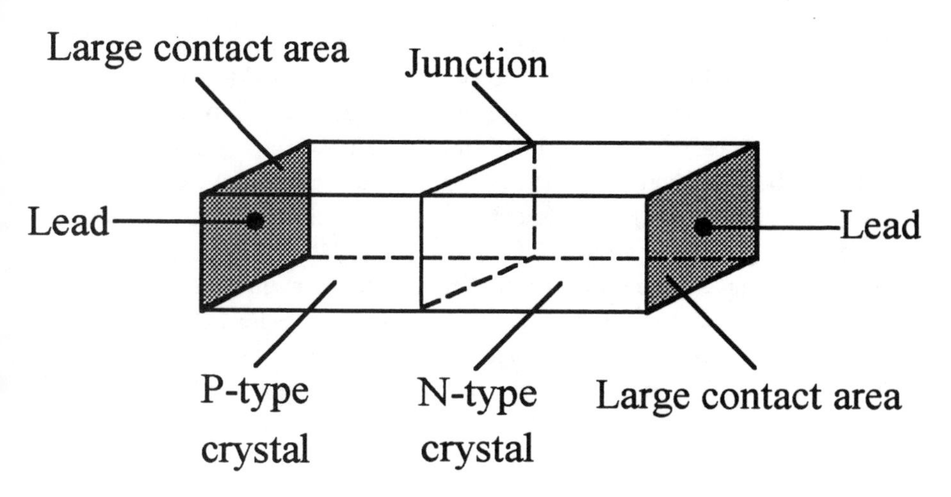

Figure 10-9. Plan of a semiconductor diode.

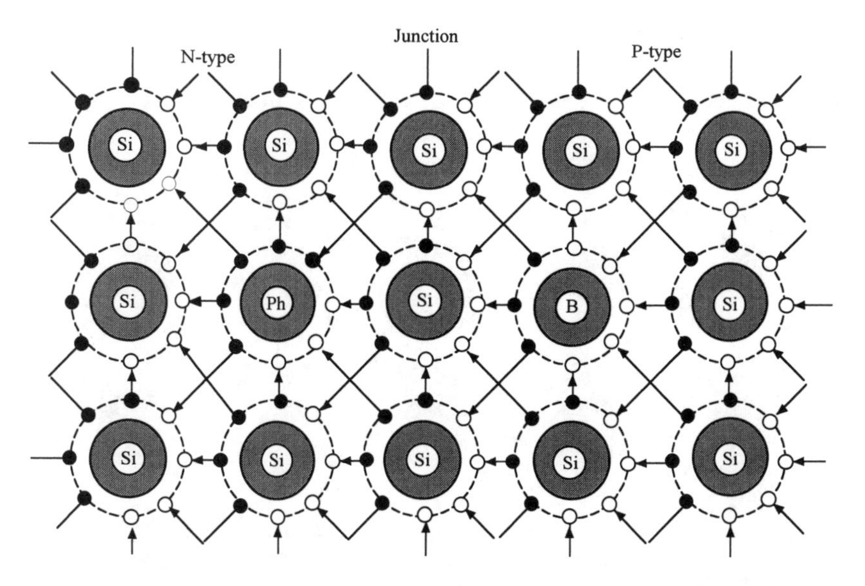

Figure 10-10. Atomic arrangement in a junction diode.

Now, let us follow the *forward conduction* of electricity through the junction diode. In Figure 10-11, a battery is connected to a diode; the negative polarity is applied to the N-type material, and the positive polarity is applied to the P-type material. This causes current through the diode. Let us consider the path of several current carriers through the diode. In the forward-conduction action, the negative pole of the battery supplies the electrons to the N-type material. The electron is passed along to the phosphorus atom, which normally has five outer electrons. At the same time, the battery takes an electron from the P-type material. This electron is supplied by the boron atom, which normally has three outer electrons. Finally, the boron atom takes a fourth electron supplied by the phosphorus atom, because of the covalent forces in the crystal. This process is repeated as long as the battery is connected. Remember, there are billions of silicon atoms and millions of impurity atoms in the crystal.

A very small forward voltage across a junction causes a large forward current. This means that the forward resistance of a diode is near very low. On the other hand, a large reverse voltage across a diode results in a minute current. This means that the reverse resistance of a diode is near infinity.

Example 10-2

Consider the method for testing a diode in Figure 10-12. In figure 10-12(a) the ohmmeter is connected across the diode for a forward resistance measurement. The connection of a VOM will result in a typical for-

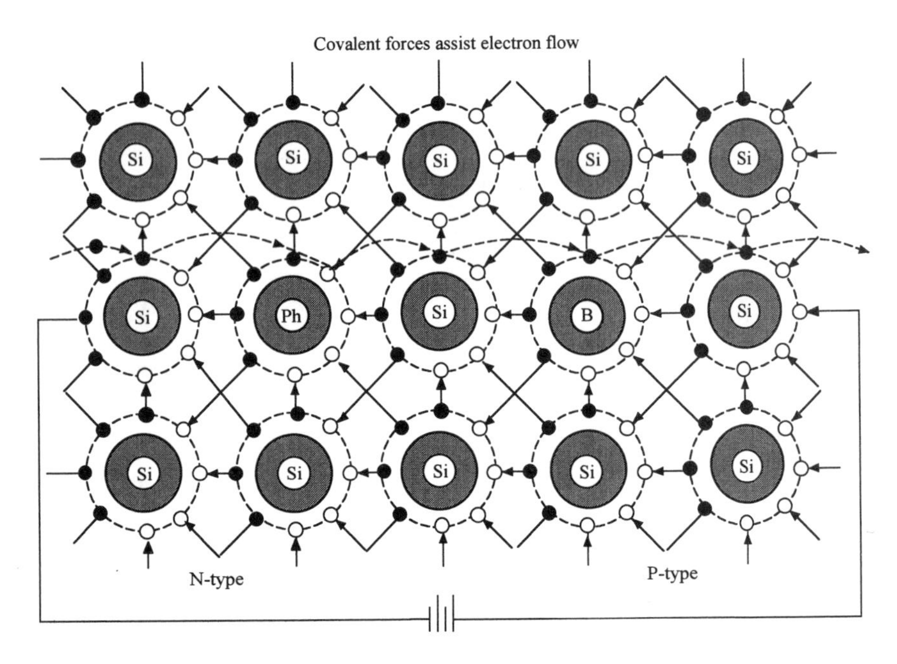

Figure 10-11. Forward conduction through a junction diode.

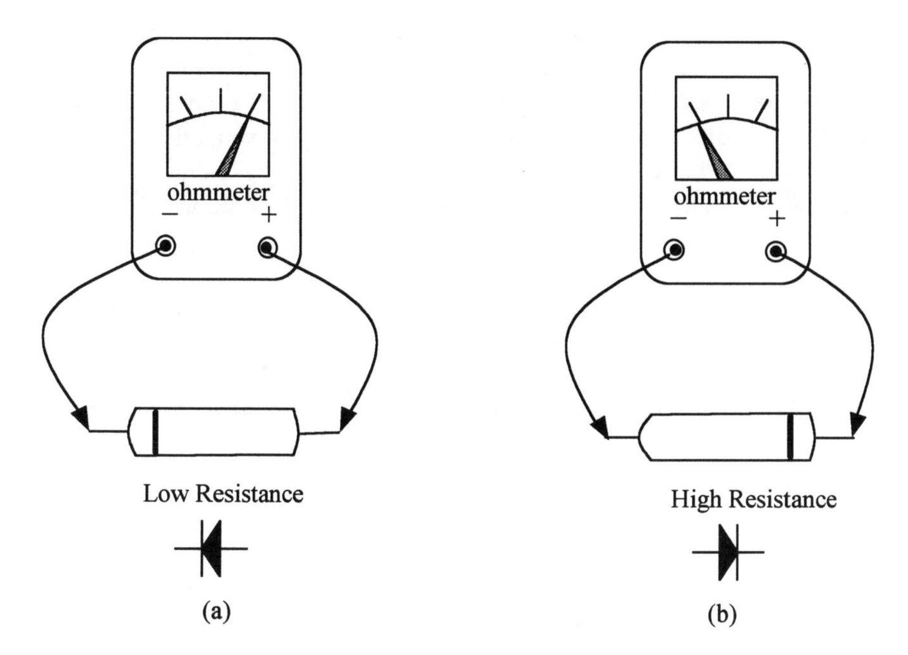

Figure 10-12. Testing the front-to-back resistance of a PN junction diode.

ward resistance reading of 5 to 50 ohms. *A DVM has a special diode function that reads the forward junction voltage of the diode; usually 0.2 V to 0.3 V for germanium and 0.5 V to 0.7 V for silicon.*

In Figure 9-12, the ohmmeter is connected to measure the reverse resistance of a diode. This measurement with either a DVM or VOM should be very high, perhaps as much as 100 Meg ohms. The ratio of these two measurements is called the front-to-back ratio, and should never be less than 1:100 for a silicon diode and never less than 1:10 for a germanium diode.

Diodes, as all electrical and electronic components are rated in several ways. Diodes are rated by:

- A maximum temperature. Silicon at 185 degrees C and Germanium at 85 degrees C.
- A maximum forward current rating.
- A maximum reverse voltage rating.

There are thousands of types of silicon diodes. Each type is identified by a number and a letter. For example: 1N150. This number may be written on the case of the diode or may be identified by the color code with bands of brown, green and black near the negative (cathode) end of the diode. A diode should be replaced with one of the same type when possible. However, any diode can be replaced with another that is *of the same type semiconductor and has higher rating in all respects.*

THE JUNCTION TRANSISTOR

The basic transistor consists of a semiconductor with two diode junctions, as shown in Figure 10-13. Notice that the transistor can be considered as a slab of semiconductor sandwiched between two blocks of opposite-type material. Thus, we have PNP and NPN arrangements. We will find that a PNP transistor has the same action as an NPN transistor. The only difference is the polarity of the power source in the circuit. The elements of a transistor are the emitter, the base and the collector. The junction between the emitter and the base is called the *emitter-base* junction; and the junction between the base and the collector is called the *collector-base junction*. In most circuits, the emitter-base junction is forward biased and the collector-base junction is reverse biased.

Combining the two diode junctions into the circuit in Figure 10-14 results in a surprising fact that a very small current in the base results in a large current in the emitter. Three factors cause this:

- the emitter is very heavily doped to supply many current carriers,
- the base is very lightly doped to receive only a few current carriers,
- the base region is very thin.

These factors cause the heavy current from the emitter to drift through the base into the collector to become output current. The ratio of the collector output current to the base input current can be as much as 300:1. This ratio is called the forward current gain beta (β).

$$B = I_c / I_B \tag{10.2}$$

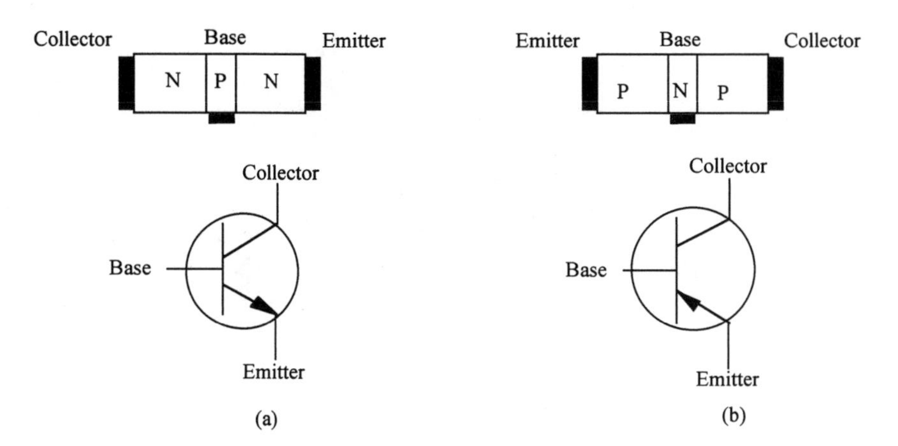

Figure 10-13. Basic transistor arrangements: (a) an NPN transistor and circuit symbol, (b) NPN transistor and circuit symbol.

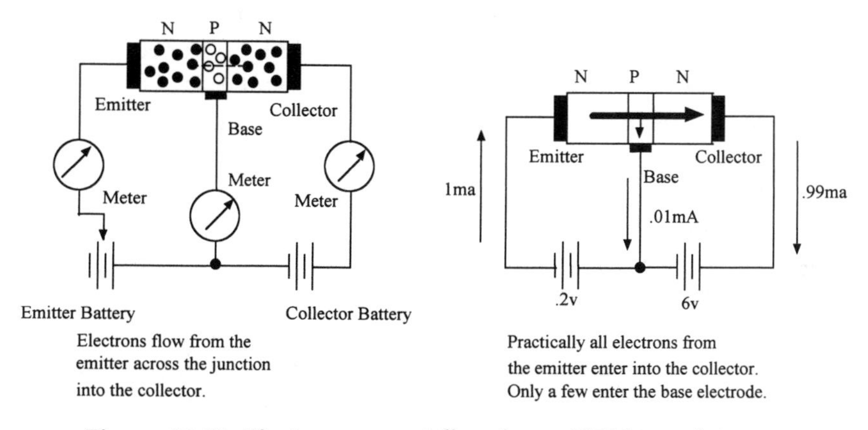

Figure 10-14. Electron current flow in an NPN transistor.

TRANSISTOR AMPLIFIERS

The common-emitter circuit (Figure 10-15) is the most widely use type of transistor amplifier because it has the highest power gain.

Example 10-3

Figure 10-15 represents a common-emitter amplifier. Increasing the resistance R_2 results in the increase of base voltage; and therefore, base

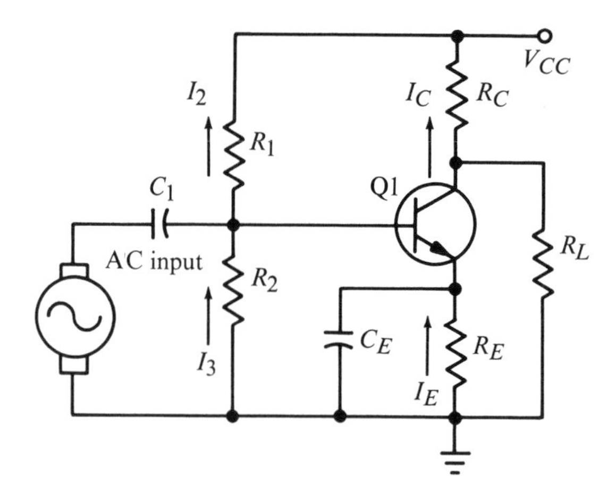

Figure 10-15. Common-emitter amplifier.

current. The increase in base voltage results in an increase of collector current and a resulting decrease in collector voltage. Suppose that β = 100; what is the increase in collector current when the increase in base current is 10 μA? What is the change of collector voltage if the collector resistor is 5 kΩ?

$$I_c = BI_B = 100 \times 10\ \mu A = 1mA$$

$$\Delta V = 1mA \times 5k\Omega = 5V$$

The change of collector current is 1 mA, which causes a decrease in collector voltage of 5 volts.

LIGHT-EMITTING DIODE

The light-emitting diode (LED) (Figure 10-16) is used primarily as an indicating lamp. The arrow out of the LED symbol indicates that light is emitted from the diode. LEDs may be purchased in numerous colors, such as white, green, yellow, red, or blue. The LED is similar in many respects to the junction diode. For example, each conducts current in only one direction, and each has a high front-to-back resistance ratio. However, the front voltage drop of an LED is approximately 1.7 V, depending upon the color. The silicon junction diode has a forward voltage drop of 0.5 to 0.7 volts. The LED is not to be used as a rectifier.

The advantage of the LED as an indicator is that it requires a low voltage and very little current, as shown in Figure 10-17.

Care must be taken in replacing an LED to assure that it is installed correctly in the circuit. The reversal of an LED in a circuit will result in a burnout. The cathode of an LED is located on the flat side of the case as shown in Figure 10-16.

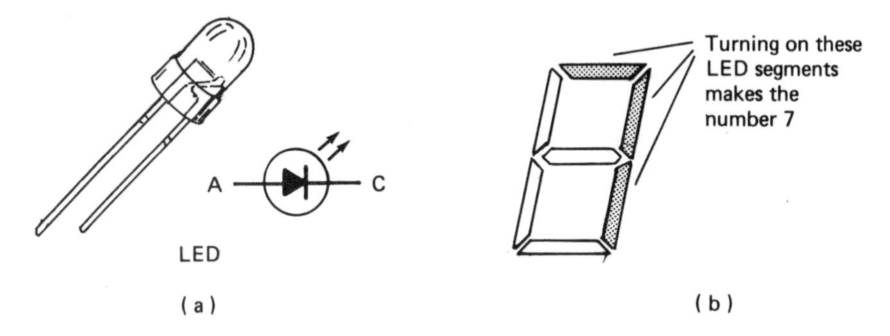

Figure 10-16. Light-emitting diode (LED): (appearance, (b) schematic symbol.

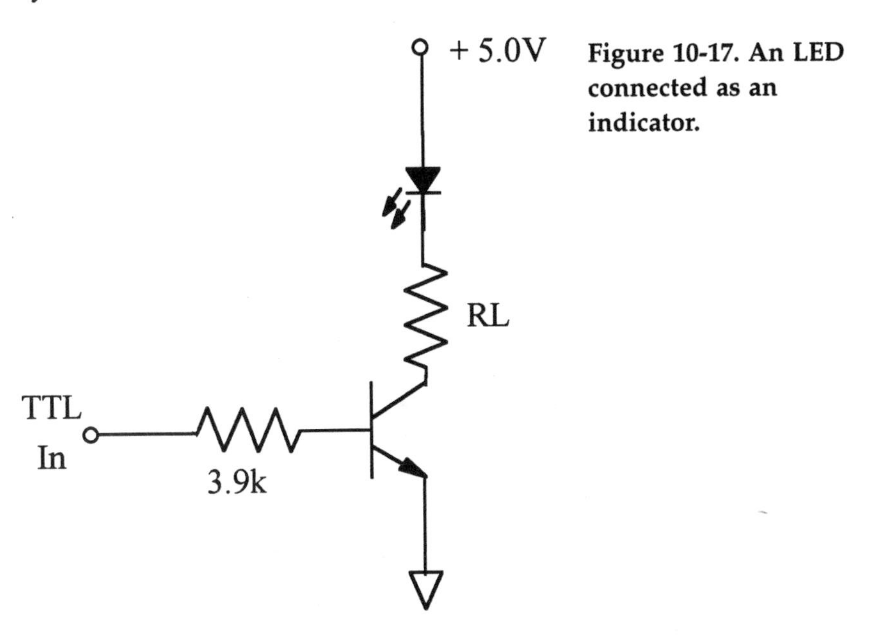

Figure 10-17. An LED connected as an indicator.

THE PHOTO DIODE

The photo diode acts as a light-controlled variable resistor. When a light is shined into the junction through the lens, electrons are freed. This reduces the resistance of the diode and allows more current in a circuit. The photo diode can be used to convert light into electrical responses. For example, it could be used to furnish indication of the presence of a pilot light.

THE PHOTO TRANSISTOR

The photo transistor shown in Figure 10-18 operates like a transistor, except the base input is light energy instead of electrical energy. When a light is beamed into the base junction, carriers are freed to become collector current. The photo transistor can be used the same as a photo diode; however, it has the advantage of greater response to light. The PT will have a response of beta times that of the PD. This results in greater gain that can be used to control circuits.

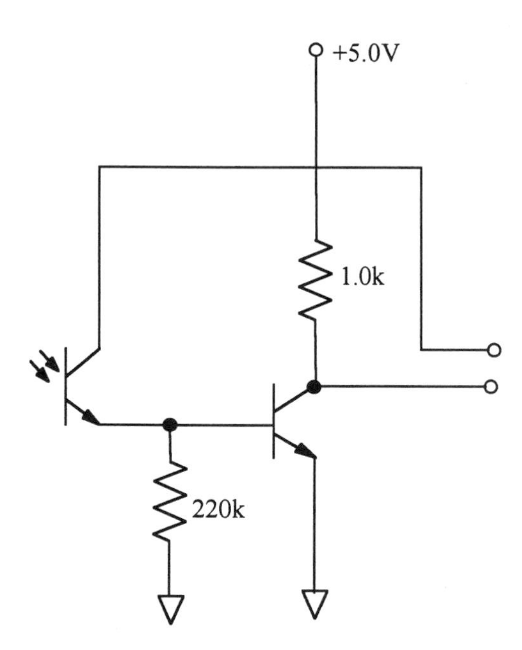

Figure 10-18. A photo transistor circuit.

THE OPTICAL-COUPLER ISOLATOR

The optical-coupler isolator, Figure 10-19, is an LED and a photo transistor packaged in one unit. The input to the device is a signal voltage to the LED that varies the light output. The light variations from the LED are picked up by the optical lenses of the photo transistor, and converted to output current in the collector.

The optical isolator operates in electronic control units as a solid-state relay that connects the CPU to the load. The output current rating of an OP ranges from 0.1 A to 5 A, depending upon the type. The device is used as an output from a central microprocessor (computer) as the control to a power transistor, a relay, or a triac, depending upon the load requirements. The triac is usually used when an AC load is to be controlled.

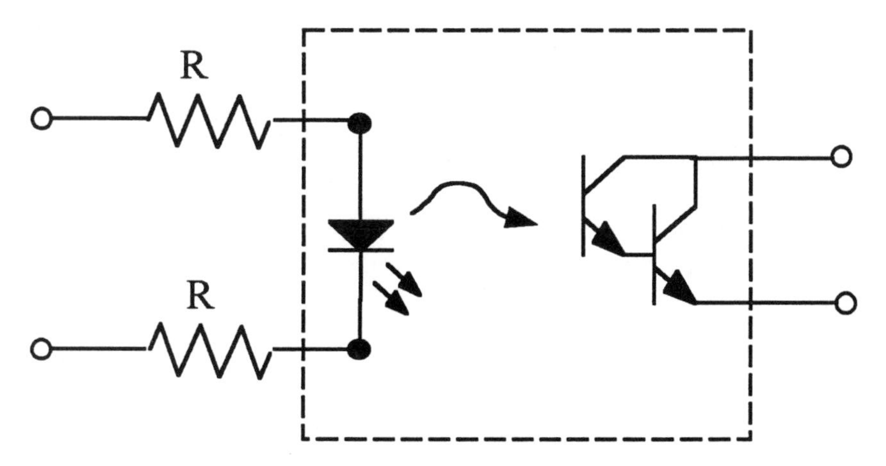

Figure 10-19. An optical-coupler isolator.

THE UNIJUNCTION TRANSISTOR

The unijunction transistor (UJT) shown in Figure 10-20, is a semiconductor device which has electrical characteristics quite different from those of a bipolar transistor. The basic characteristics of a UJT are:

1. The basic elements B_1 and B_2 are open (high resistance) until the emitter-base voltage is raised to a positive threshold value.

2. The emitter to B_1 appears open until the threshold voltage is reached,

3. The emitter-to-base resistance approaches zero when the threshold voltage is reached.

4. At the threshold voltage the resistance between B_1 and B_2 approaches zero.

When a capacitor and resistor are placed in the emitter as shown in Figure 10-21, a free running oscillator is formed. The capacitor begins to charge to V_{BB}, since the emitter is open. When the voltage on the capacitor reaches the emitter threshold voltage it is shorted across the emitter to base B_1. The capacitor quickly discharges and again begins to charge to $V_{BB.}$ The repeated charging action of the capacitor produces a sawtooth voltage, resulting in a free-running oscillator. The output of B_1 or B_2may be used to trigger other circuit actions. The circuit finds application in electrical controllers.

THE SILICON CONTROLLED RECTIFIER

The silicon controlled rectifier (SCR) is a three-terminal device that is normally open until the appropriate signal voltage is applied to the

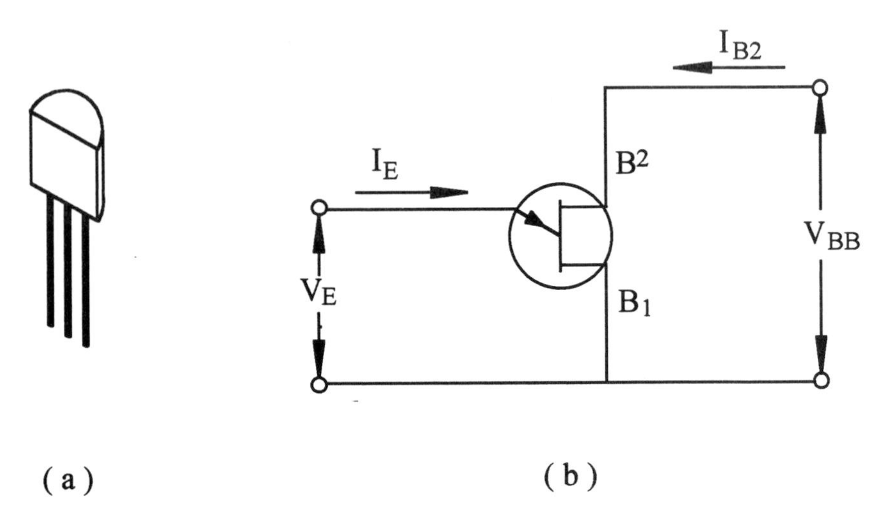

Figure 10-20. Unijunction transistor: a. appearances, b. symbol.

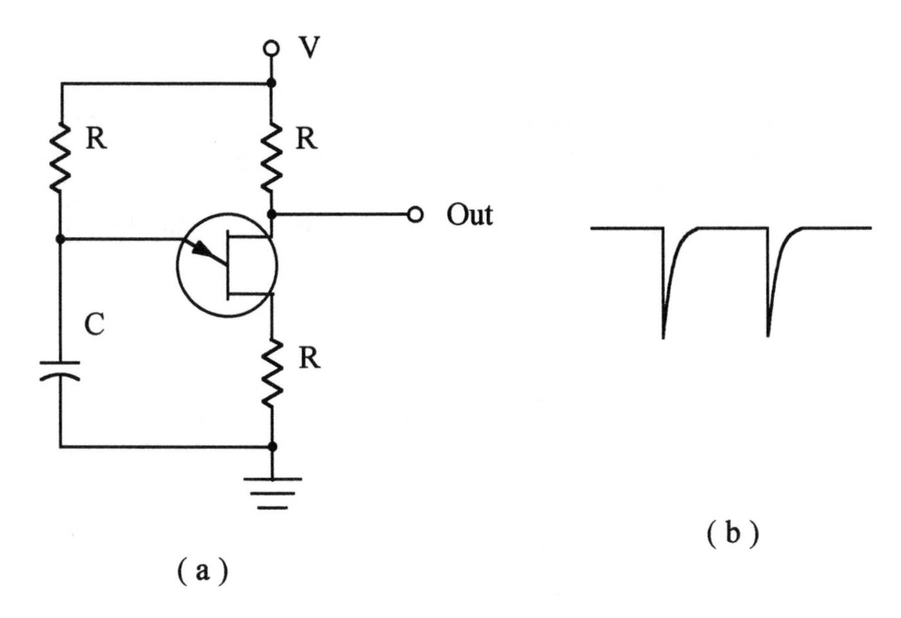

Figure 10-21. A circuit using a unijunction transistor.

gate terminal, at which time it rapidly switches to the conducting state. Figure 10-22 shows the appearance of a typical SCR with the standard symbol and termination designation. The SCR may be regarded as a modified rectifier diode, which operates as an electronic switch with no moving parts. The conduction state of the SCR is controlled by a trigger voltage applied to the gate. The SCR remains turned on after the trigger voltage is removed.

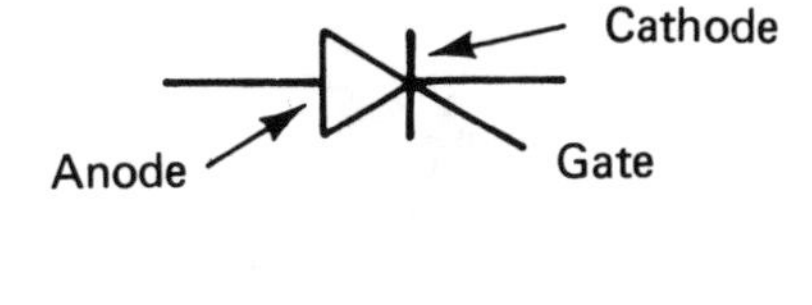

Figure 10-22. Silicon Controlled rectifier: (a) appearance. (b) symbol, (c) symbol.

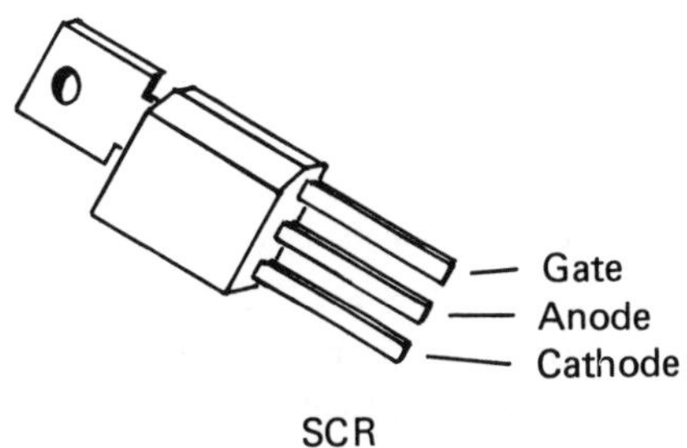

THE DIAC

The DIAC is a bi-direction switch that is normally an open circuit until a specific level of voltage of either polarity is applied. A that point the DIAC starts conducting. Therefore, the device is regarded as a solid-state switch. The DIAC (Figure 10-23) goes into the non-conducting state every time the voltage goes to zero.

THE TRIAC

A TRIAC is a three-terminal AC switch that has the same characteristics as an SCR, except that it will conduct in either direction. Its appearance and symbol are shown in Figure 10-24. Like the SCR, the DIAC and TRIAC are useful in circuits that control large amounts of current. Figure 10-25 shows an example of a TRIAC in a motor speed control circuit. The voltage setting on potentiometer R_1 controls the conduction of the TRIACs, which controls the current in the AC motor.

THE OPERATIONAL AMPLIFIER

The operational amplifier (OP AMP) is another solid-state device that is becoming common in electronic control systems. The operational amplifier appears much like a many-lead transistor. However, a typical

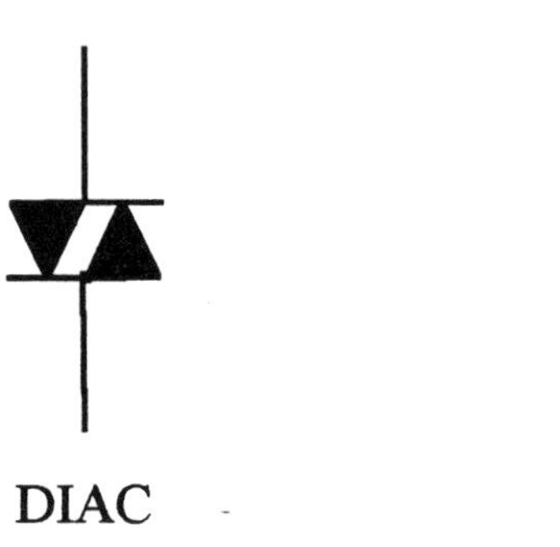

Figure 10-23. Diac, or bi-directional switch: (a) symbol.

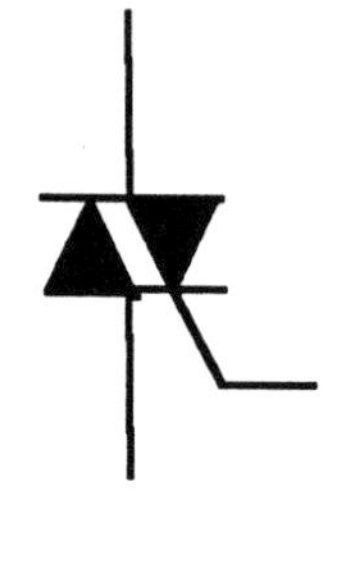

Figure 10-24. Circuit symbol of a TRIAC, or three-terminal bi-directional switch.

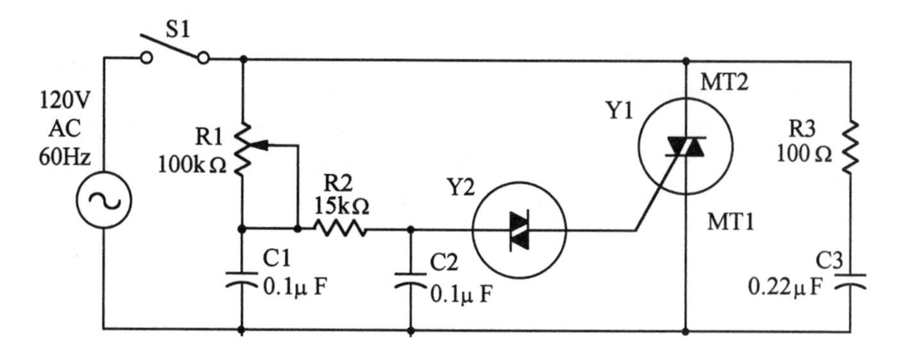

Figure 10-25. A TRIAC as a motor controller.

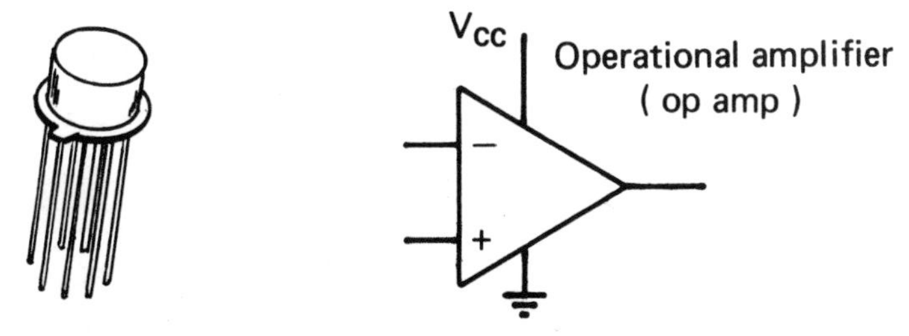

Figure 10-26. Operational amplifier appearance and circuit symbol.

OP AMP contains perhaps 20 transistors, 6 diodes and 12 resistors. The combination of these devices in a very high-gain circuit into one small package makes it one of the most versatile of semiconductor devices. The OP AMP, shown in Figure 10-26, has a gain of 200,000 and will amplify both DC and AC voltages. The unit is packaged in both a flat pack and a round pack. A package may contain more than one OP AMP.

THERMISTOR

The thermistor is a two-terminal semiconductor that may have either a positive or a negative temperature coefficient. That is, the resistance of a positive-temperature coefficient thermistor rises and falls directly with a change of temperature. On the other hand, the resistance changes inversely with temperature for a negative-temperature coefficient thermistor.

A positive-temperature coefficient thermistor can be used as a solid-state motor starting relay (Figure 10-27). The thermistor is connected in series with the starting winding. The cold resistance is very low, 2 to 6 ohms, which allows a large current to flow through the starting winding, developing a large torque. As the thermistor heats, its resistance increases, reducing the winding current.

TROUBLESHOOTING

The testing techniques discussed here use the ohmmeter on a VOM or a digital VOM. The authors suggest that a DVM be part of your tool kit because of its rugged construction. The OHM selection on the DVM should be used for all these tests. The tests involve making a front-to-back ratio test across a PN junction as we did for the junction diode and junction transistor in Chapter 9.

Testing Diodes

The majority of diode and transistor faults can be detected by the use of the ohmmeter on a VOM or the diode test on a DVM to establish the front-to-back ratio of a PN junction. Note, the ohmmeter function on a DVM does not have enough voltage to cause forward conduction on a silicon PN junction.

Figure 10-28 illustrates the method of using the ohmmeter to check the front-to-back ratio of a PN junction, in this case that of a diode.

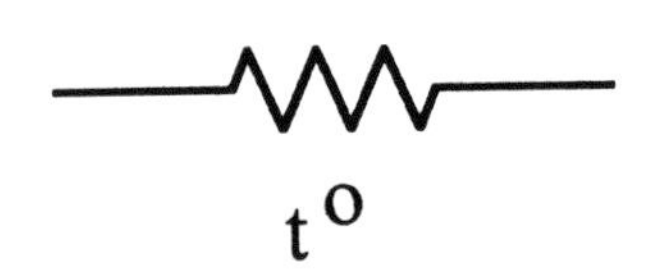

Figure 10-27. The thermistor circuit symbol.

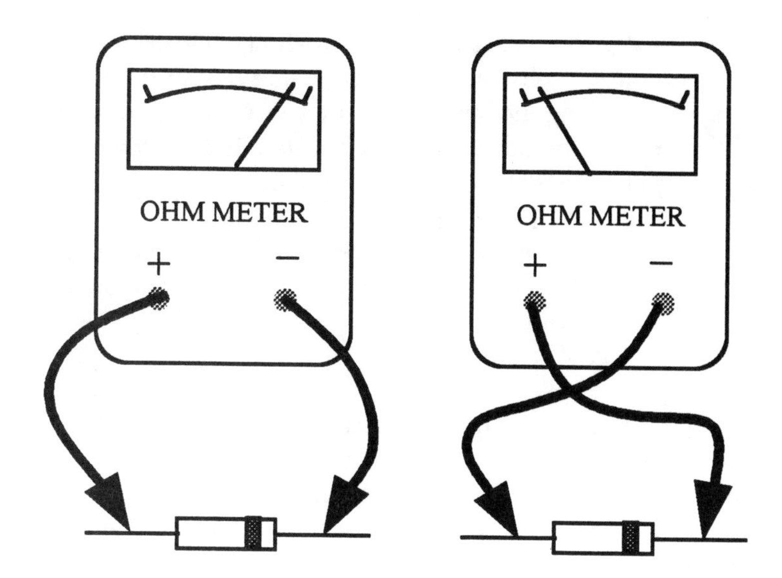

Figure 10-28. Using an ohmmeter to test a PN junction diode.

- The forward resistance is measured in Figure 10-26(a). An ohmmeter on a VOM turns on the PN junction causing current and a resulting low resistance reading. A reading of 20 to 100 ohms may be expected, depending upon the scale selected on the meter. The DVM diode scale reads the forward junction voltage of the diode, a value of 0.5 to 0.7 volts. Any other reading indicates a shorted or open diode.

- Figure 10-28(b) illustrates the method for measuring the reverse resistance of a PN junction diode, again that of a junction diode. Both the VOM and the DVM should read infinity or at least 60 Meg ohms. A low reading indicates a short or excessive leakage. The ratio of the reading is called the front-to-back ratio. This should be well over 100 to 1.

Testing Transistors

Remember from the discussion on semiconductor theory that a transistor has two PN junction—one between the base and the emitter and another between the collector and the base. These junctions are tested in the same manner as was that of a diode, Figure 10-29.

- Figure 10-29(a) illustrates the correct method of testing the forward resistance of the collector-base junction and the emitter-base junction. The resulting measurement with a VOM and a DVM should be the same as that for the junction diode. That is a low value for the VOM and 0.5 to 0.7 volts for the DVM.

- Figure 10-29(b) illustrates the correct method for measuring the forward resistance of the transistor junctions. The reading should be above several hundred Meg ohms. Again, the front-to-back ratio should be well over 100 to 1.

- The emitter-to-collector and collector-to-emitter of the transistor can also be tested with an ohmmeter. With the negative lead on the emitter and the positive lead on the collector a reading of around 20 k to 100 k ohms should be observed on the VOM. The DVM on the OHMs scale should indicate open. With the negative lead on the collector and the positive lead on the emitter, higher reading should be indicated on the VOM and an open on the DVM.

A very low reading in either direction indicates a short circuit. Figure 10-30 gives relative resistance values of the foregoing measurements. The measurements will vary from transistor to transistor. However, their relationship should be similar.

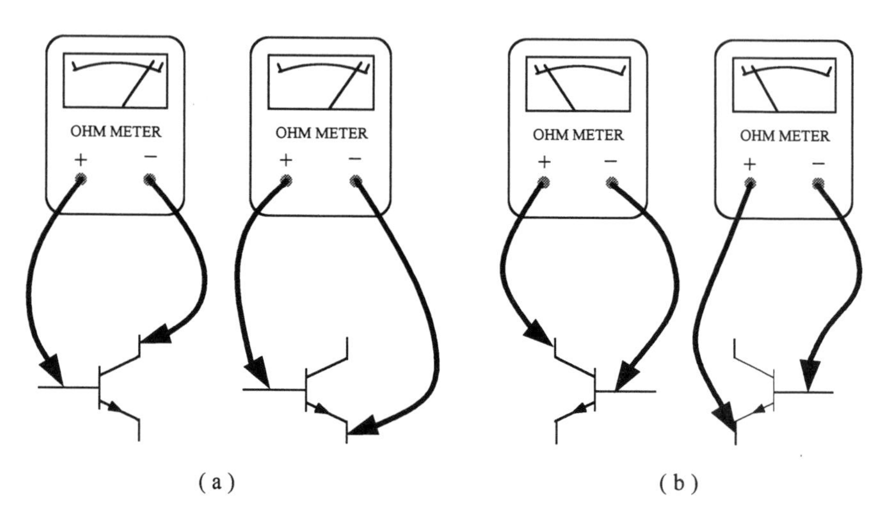

Figure 10-29. Testing a transistor with an ohmmeter.

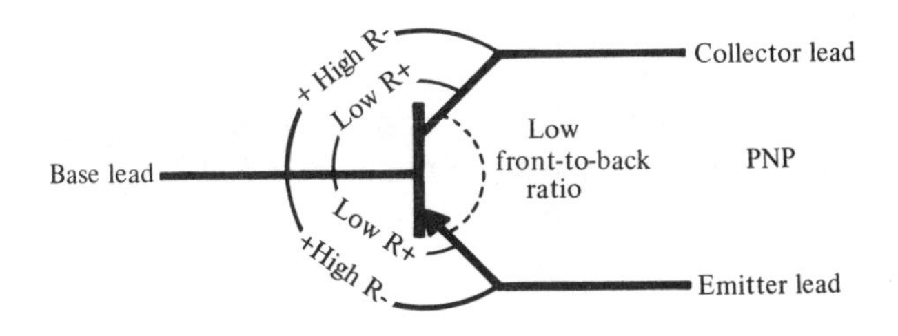

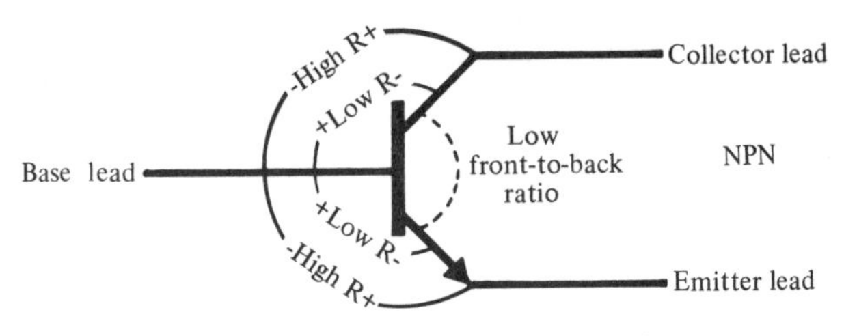

Figure 10-30. Relative resistance values of an ohmmeter test on a transistor.

LED

The LED is a diode and can be tested for a front-to-back ratio, as we did the junction diode. The front-to-back ratio should measure more than 100 to 1.

Photo Diode

The photo diode is tested, for a front-to-back ratio, in the same manner as the junction diode and the LED. The ratio should be similar. The variation of the resistance can be observed by shining a light into the junction lens.

Photo Transistor

The junctions of the collector-base and the collector-emitter junctions of the photo transistor cannot be accessed for measurements as was the junction transistor. However, a resistance measurement can be made between the emitter and the collector. The positive lead is connected to the collector and the negative lead is connected to the emitter. A reading of several thousand ohms is normal. Shining a light into the lens of the transistor should reduce the resistance reading by a large amount. The amount will depend upon the brightness of the light.

Optic-Coupler Isolator

The input to an optic coupler is an LED, and can be tested in the same manner as that device. The output of the optic couplet is a photo transistor. It can be tested in the same manner as that device. However, be aware that you are testing the entire device when shining a light into the input.

Unijunction Transistor

The unijunction transistor can be tested with an ohmmeter as shown in Figure 10-31. The resistance between B_1 and B_1 should measure a few hundred ohms regardless of the lead connections (+ to – or – to +) A diode front-to-back ratio test can be made between the emitter and B_1 or the emitter and B_2.

Silicon Controlled Rectifier

Three tests must be performed to assure that an SCR is operational.

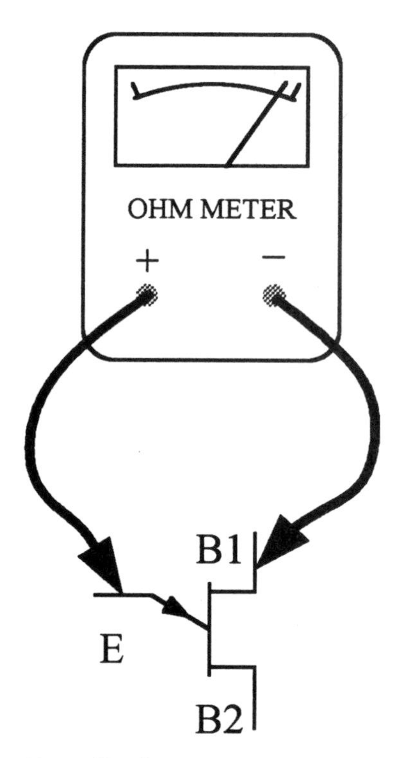

Figure 10-31. Testing a UJT with an ohmmeter.

1. Place the positive lead of the ohmmeter on the anode and the negative lead on the cathode. The meter should read an open or infinity, Figure 10-32(a).

2. Connect the gate to the anode. The SCR should turn on and the meter should read a short, Figure 10-32(b).

3. Connect the positive lead to the cathode and the negative lead to the anode. The meter should read an open, Figure 10-32(c).

4. Connect the gate to the anode. The meter should again read an open. A high current SCR may read a very high resistance less than infinity due to leakage current, Figure 10-32(d).

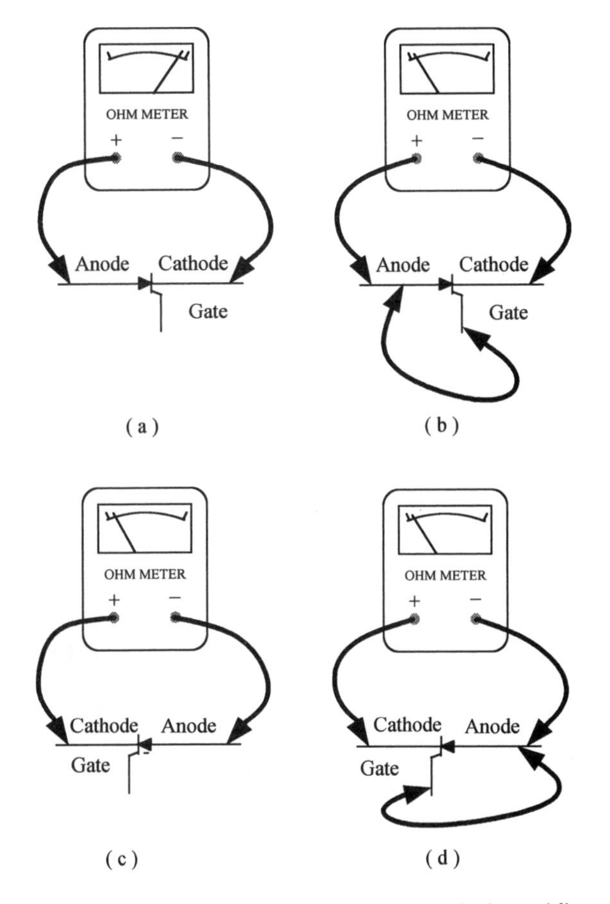

Figure 10-32. Testing a silicon controlled rectifier.

DIAC

The DIAC can be tested with an ohmmeter, a voltmeter and a power supply or 24-V AC source.

1. Connecting the ohmmeter in either direction should result in an open reading, because the DIAC requires a voltage greater than that of the ohmmeter to turn on.

2. Connect the DIAC in the circuit shown in Figure 10-33 with the voltmeter connected across the resistor. Apply the source voltage. The Diac should turn on, and the resulting current should cause a voltage drop across the resistor.

TRIAC

The TRIAC is tested like the SCR. The technique is shown in Figure 10-34.

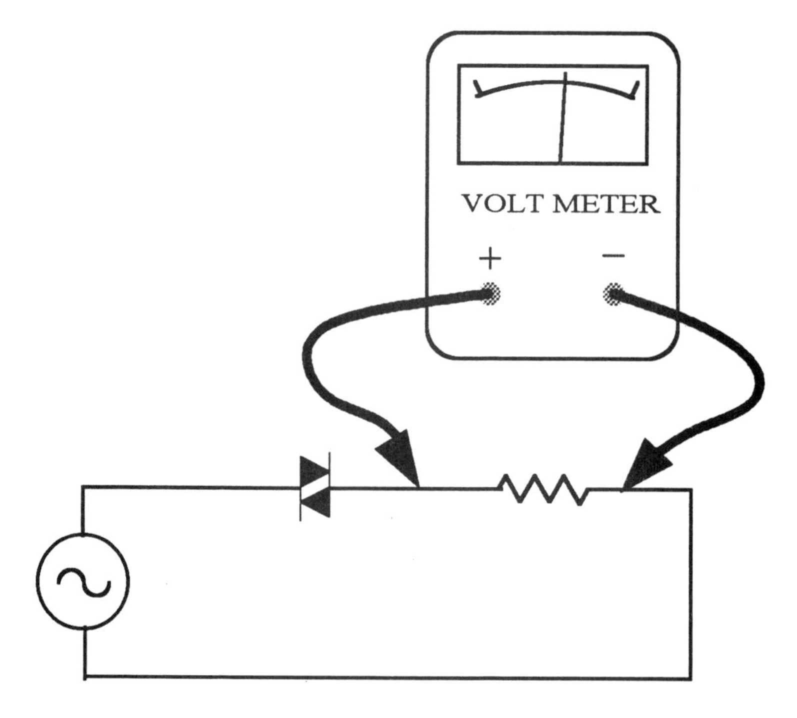

Figure 10-33. Testing a DIAC.

a. Connect an ohmmeter across the TRIAC. The meter should read infinity or an open.

b. Jumper the gate to the negative lead of the meter. The meter should read a low resistance.

c. Reverse the meter connection to the connection and repeat steps 1 and 2. The results should be the same.

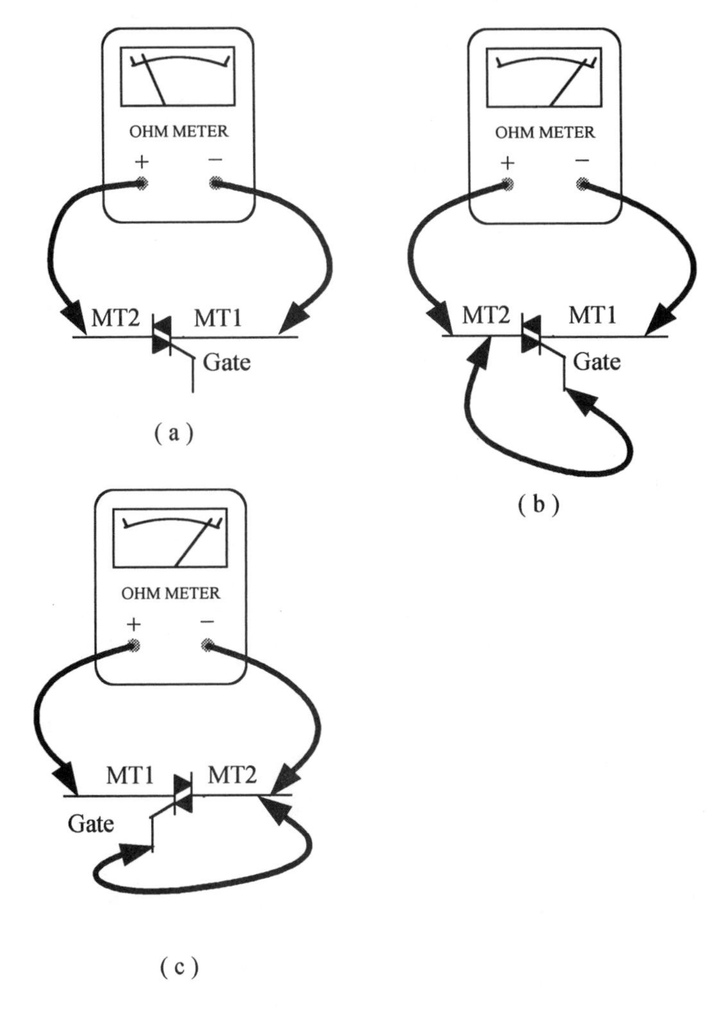

Figure 10-34. Testing a TRIAC.

Operational Amplifier

The OP AMP cannot be tested effectively out of the circuit. An in-circuit test can be performed. However, a new device usually must be installed for a final determination.

- With the power applied a voltage test is made at the output terminal to ground. If the voltage is close to either the positive or negative supply the output of the OP AMP is shorted or the input bias voltage is too high.

SUMMARY

- The most commonly used semiconductor is silicon.
- The PN junction diode acts as a rectifier.
- The transistor can be used to amplify a signal, shape a waveform or generate a waveform.
- Diodes and transistors can be tested with an ohmmeter.
- The PN junction diode acts as a rectifier.
- Most semiconductor devices can be tested with an ohmmeter.
- The diode function must be used on the DVM for a PN junction test.
- The front-to-back ratio of a silicon junction should be more than 100 to 1.

Chapter 11

Electric Control Devices And Circuits

INTRODUCTION

Control circuits can be grouped into various categories such as motor on-off controls, motor speed controls, electronic timers, time-delay units, synchronous switches, heat controls, climate controls, an so on. Some control circuits' processes consume considerable power, whereas others switch minute amounts of power.

Control devices are depicted in wiring diagrams with specific circuit symbols. These symbols must be understood to follow the electrical flow and sequential function of air conditioning, refrigeration, and heating systems. Table 11-1 is a summary of switch, contactors, and relay contact configurations.

Table 11-2 depicts the symbols for switches operated by pressure, liquid level, temperature, air or liquid flow, and overload. The pressure switches are shown as typical N.O. and N.C. that are operated by either a low pressure or a high pressure. In addition the differential pressure switch is operated by both a high pressure and a low pressure within a differential limit.

The thermal operated switch symbols are shown for N.O. and N.C. contacts. Normally closed thermal switches are always safety switches.

Table 11-3 depicts the wiring symbols for switch contacts that are controlled by time, and switch contacts that are instant operating. The timing switch contacts are closed at a predetermined time after the coil is energized.

Table 11-4 depicts the additional wiring symbols that are necessary to fully understand wiring diagrams. Coils for relays or contactors appear on almost all wiring diagrams. Shunt coils are indicated by a circle

SPST, N.O.		SPST, N.C.		SPDT		TERMS
SINGLE BREAK	DOUBLE BREAK	SINGLE BREAK	DOUBLE BREAK	SINGLE BREAK	DOUBLE BREAK	SPST - SINGLE POLE SINGLE THROW
						SPDT - SINGLE POLE DOUBLE THROW
DPST, N.O.		DPST, N.C.		DPDT		DPST - DOUBLE POLE SINGLE THROW
SINGLE BREAK	DOUBLE BREAK	SINGLE BREAK	DOUBLE BREAK	SINGLE BREAK	DOUBLE BREAK	DPDT - DOUBLE POLE DOUBLE THROW
						N.O. - NORMALLY OPEN N.C. - NORMALLY CLOSED

Table 11-1. Standard contact symbols.

PRESSURE & VACUUM SWITCHES		LIQUID LEVEL SWITCH		OVERLOAD RELAYS	
N.O.	N.C.	N.O.	N.C.	THERMAL	MAGNETIC

TEMPERATURE ACTUATED SWITCH		FLOW SWITCH (AIR, WATER, ETC)	
N.O.	N.C.	N.O.	N.C.

Table 11-2. Control symbols.

CONTACTS							
INSTANT OPERATING				TIMED CONTACTS - CONTACT ACTION RETARDED AFTER COIL IS:			
WITH BLOWOUT		WITHOUT BLOWOUT		ENERGIZED		DE-ENERGIZED	
N.O.	N.C.	N.O.	N.C.	N.O.T.C.	N.C.T.O.	N.O.T.O.	N.C.T.C.

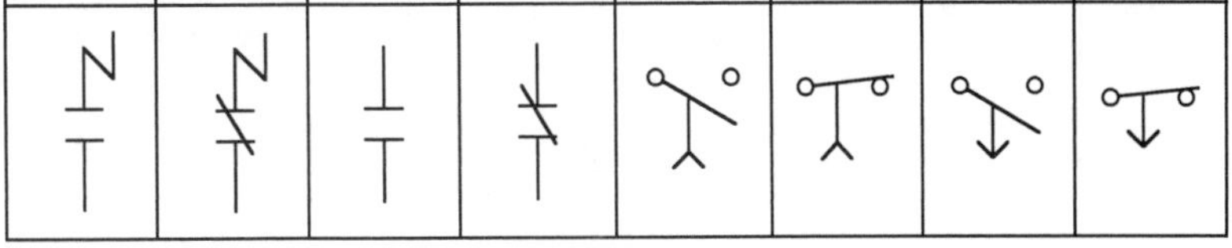

Table 11-3. Instant operating and timed switches.

and series coils are indicated by slashed lines. Inductors, motor windings and transformers appear as looped lines.

The symbols in the foregoing tables should be studied and must be referred to as we study electrical control devices.

THERMOSTATS

Thermostats are devices used to control heating and cooling. Basically, a thermostat is comprised of two parts: a temperature-sensitive element, and a control device or mechanism. Thermostats are the basic devices used to call for heat or cold in air conditioning, refrigeration, and heating.

Thermostats that control temperature within a space, such as a home or office, usually use a simple bi-metal strip as the temperature-sensing element. The bi-metal strip is usually wound into a spring to obtain greater length for improved accuracy. One electrical contact of a switch may be attached to the bi-metal spring. When the contacts close a signal is sent to the heating or cooling system. Often the switch is placed in a vacuum bulb partially filled with mercury. The mercury bulb is attached to the bi-metal spring that expands and retracts with changes of temperature. The bulb rotates with the spring and the pool of mercury shorts across the switch contacts at a selected temperature. The advantage of the mercury bulb is that the vacuum in the bulb prevents pitting of the contacts due to arcing, thereby extending the life of the thermostat.

The thermostat in Figure 11-2 is designed for a high-low temperature range to control both heating and cooling. The system also has a fan switch to control air circulation only.

TRANSFORMERS				
AUTO	IRON CORE	AIR CORE	CURRENT	DUAL VOLTAGE

A C MOTORS			
SINGLE PHASE	3 PHASE SQUIRREL CAGE	2 PHASE 4 WIRE	WOUND ROTOR

D C MOTORS			
ARMATURE	SHUNT FIELD	SERIES FIELD	COMM. OR COMPENS. FIELD
	(Show 4 loops)	(Show 3 loops)	(Show 2 loops)

PILOT LIGHTS		
INDICATE COLOR BY LETTER		
ILLUMINATED	NON PUSH - TO TEST	PUSH - TO TEST
R	A	R

COILS		INDUCTORS	
SHUNT	SERIES	IRON CORE	AIR CORE

Table 11-4. Miscellaneous wiring symbols.

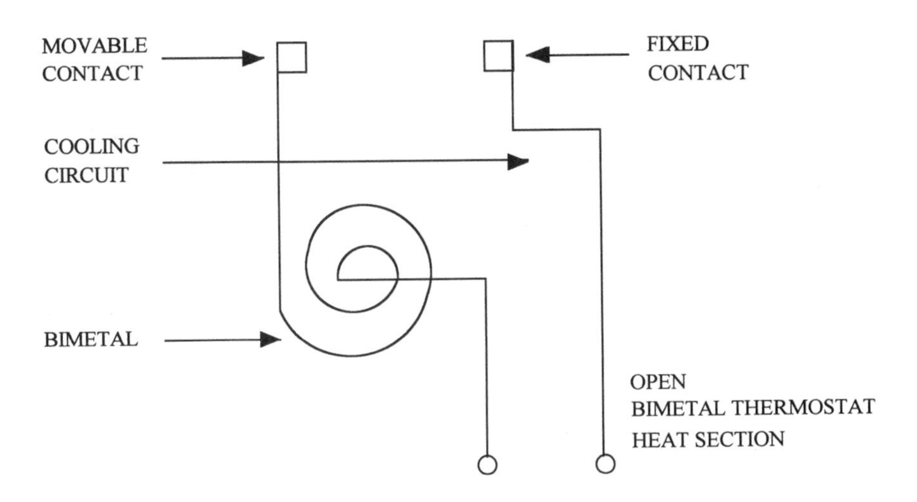

Figure 11-1. Example of a bi-metal controlled thermostat.

Figure 11-2. A thermostat to control both air conditioning and heating. *(Courtesy, Paragon Controls.)*

Some air conditioning thermostats have heating and cooling anticipators. The heating anticipator is a small resistance heater located near the bi-metal strip. This resistor preheats the bimetal control slightly and prevents overheating by the system. The thermostat turns off the furnace or heat pump several degrees before the selected temperature and continues heating as the fan blows air across the heat exchanger. The cooling anticipator operates in the reverse action of the heat anticipator when the system is in the cooling mode.

Programmable thermostats make use of microprocessors to allow selection of heating or cooling at a predetermined time. For example, a thermostat in a ski lodge might be programmed to turn on the heat before the owner's arrival on the weekend. Such systems can be controlled to respond to telephone messages. Figure 11-3 is a photograph of such a thermostat.

Solar heating systems utilize differential thermostats. The thermostat is set to differentiate between two temperatures, the temperature of the outside atmosphere, and the water temperature of the heat exchanger. For example, the solar system must be turned off before the water from a swimming pool begins to be cooled as the outside temperature approaches that of the pool water.

PRESSURE CONTROLLED SWITCHES

Pressure controlled switches are used for safety and to control temperature. Pressure controlled switches may be either N.O. or N.C. or contain both switch functions; however, all utilize a bellows for switch armature activation. Figure 11-4 depicts the basic bellows action controlled buy refrigerant pressure.

A refrigeration system may use two pressure controls. A low-pressure control is connected into the low side pressure side of the compressor. The low-pressure switch disconnects the compressor when suction pressure falls below a safe level. A low pressure often signals a loss of refrigerant which could damage the system. The high-pressure control is connected into the high-pressure side of the compressor. The high-pressure switch disconnects the compressor if the discharge pressure becomes excessive. These safety switches are N.C. and are usually connected in series with the compressor contactor coil. The high-pressure and the low-pressure controls have similar appearances; therefore, the

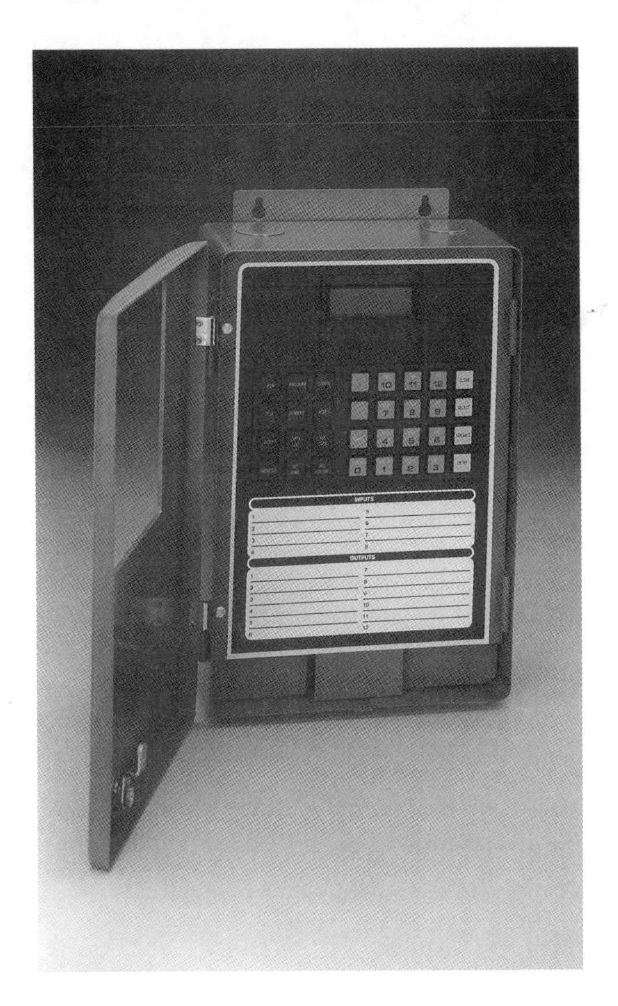

Figure 11-3. A programmable thermostat or energy management module. *(Courtesy, Paragon Controls)*

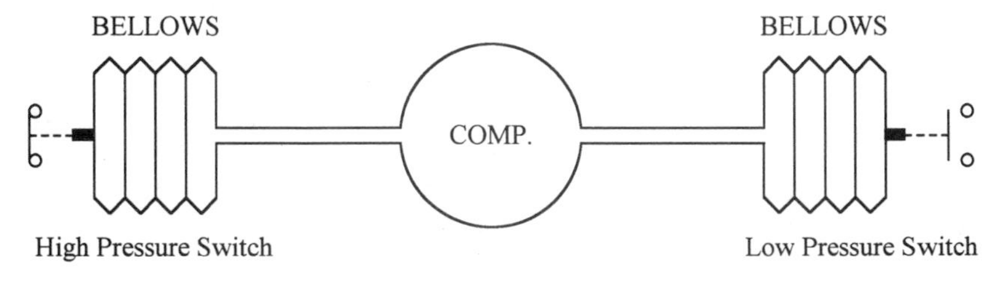

Figure 11-4. An example of a bellows switch.

technician should read the pressure information on the unit.

Figure 11-5 illustrates the application of two types of pressure switch arrangements. Figure 11-5(a) utilizes a dual action high/low pressure switch; while Figure 11-5(b) uses two separate switches. Figure 11-6 depicts a high and low pressure switch in the same package. Each switch is adjustable.

THE AIRFLOW OR SAIL SWITCH

Sail or airflow switches used in air conditioning systems operate on a flow of air rather than a flow of liquid. The sail switch, Figure 11-7, is a micro switch with a piece of light metal connected to the activating arm. When the system fan forces airflow to pass over the sail the pressure causes the arm of the micro switch to activate. This signals the system control unit that the compressor and evaporator fan motors are operating, and that the compressor can safely operate.

THE HUMIDISTAT

Humidistats are used to control humidity and moisture in some environmental control systems. It is important in many industrial environments to keep a constant humidity. It is also often desirable to keep a constant humidity in homes and offices. The humidistat operates to con-

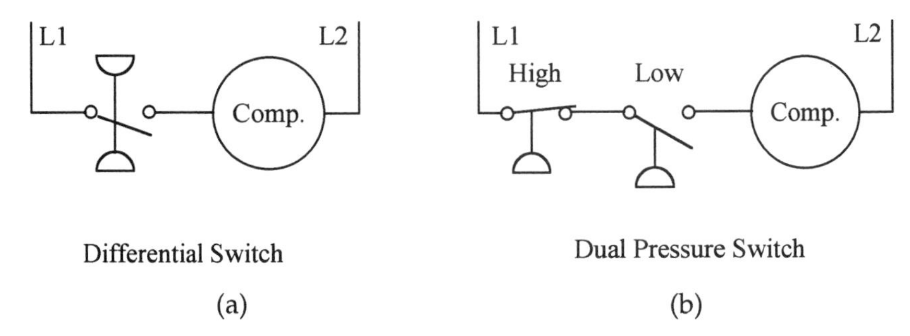

Figure 11-5. Wiring diagram of pressure switch applications: (a) single high/low switch, (b) separate high and low switches.

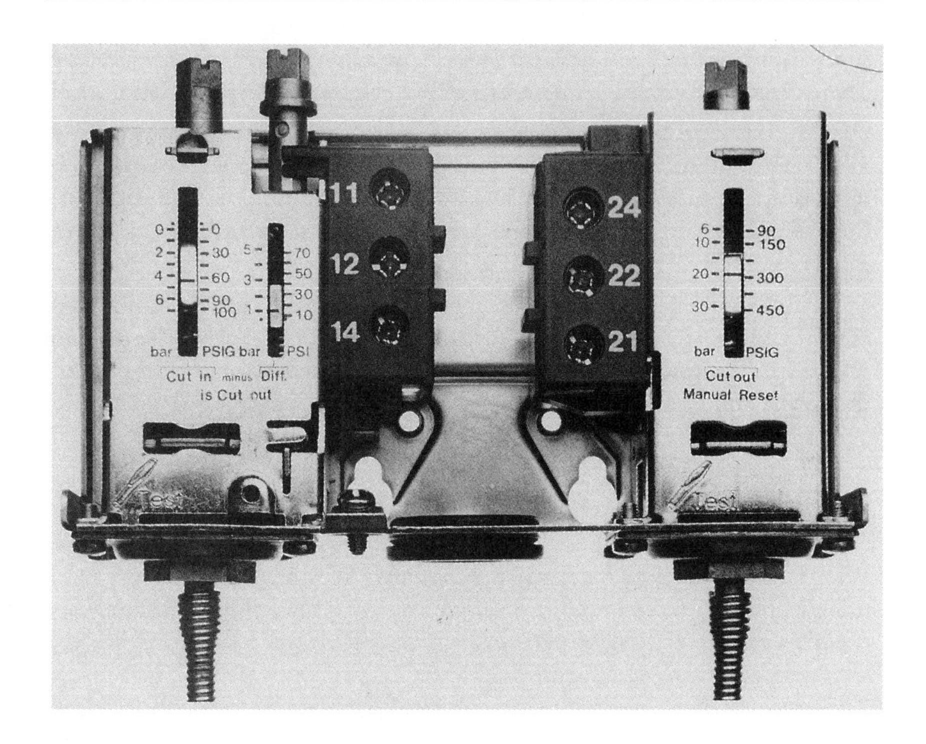

Figure 11-6. High and low pressure switch in one package. *(Courtesy, Paragon Controls)*

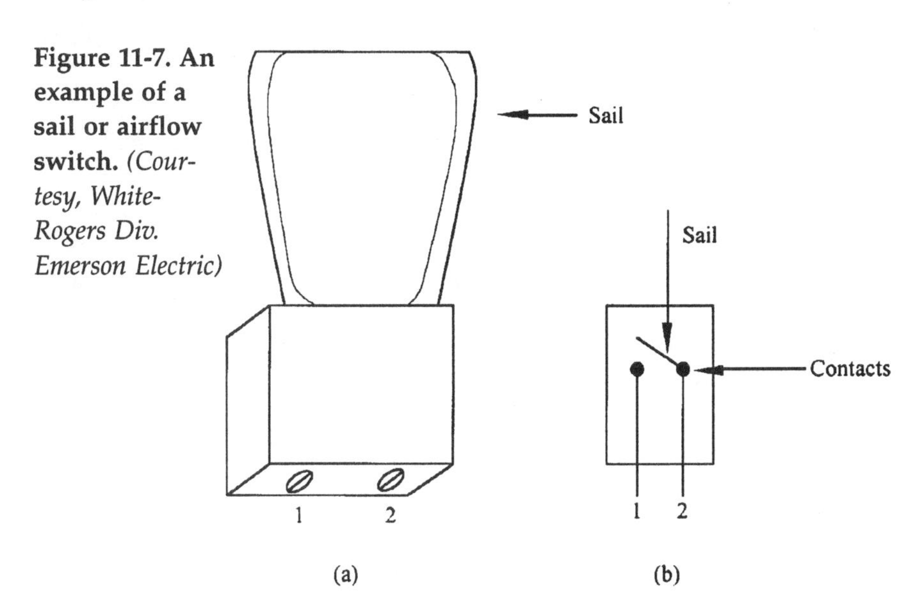

Figure 11-7. An example of a sail or airflow switch. *(Courtesy, White-Rogers Div. Emerson Electric)*

trol humidity much as the thermostat operates to control temperature. The humidistat contains an element that senses the level of moisture in airflow in a central heating and air conditioning system. It then activates a humidifier that adds water to the airflow. At first glance the humidistat, Figure 11-8, may be mistaken for a thermostat. The humidifier for the system may be located in the ducting of a heating and air conditioning system.

Most area air conditioning and heating systems employ duct-installed humidifiers to control the humidity of the air. These devices are located in the ducting system and require frequent cleaning to prevent the formation of bacteria or mold in the moisture reservoir.

TIMERS

Timers are clock mechanisms that are used to control several functions in heating and air conditioning systems. Older timing devices utilized clock mechanisms. However, modern timing devices are comprised of semiconductor components.

The *seven-day timer/thermostat,* depicted in Figure 11-9, is used to control the temperature in spaces such as offices and schools where the daily and hourly needs change. The timer consists of a microprocessor (small computer) and a cooling and heating thermostat. It can be preset to turn on and off the thermostat on certain days of the week and at definite times of the day.

Repairing the internal circuit operation of a timer/thermostat is generally beyond the ability of a field technician. Therefore, a faulty unit is usually replaced. However, it is important that the technician be sure that a unit is faulty before making a replacement. For this reason manufactures usually supply a test flow chart that can be used to evaluate the thermostat. A troubleshooting procedure is presented at the end of the chapter.

A second timer used in refrigeration and air conditioning systems is the *short-cycling timer.* Short cycling is the act of the compressor being turned off and on too soon. This action can be caused by a number of events that can occur in a system. For example, turning on and off a thermostat, low line voltage, improper phase of line voltage, momentary

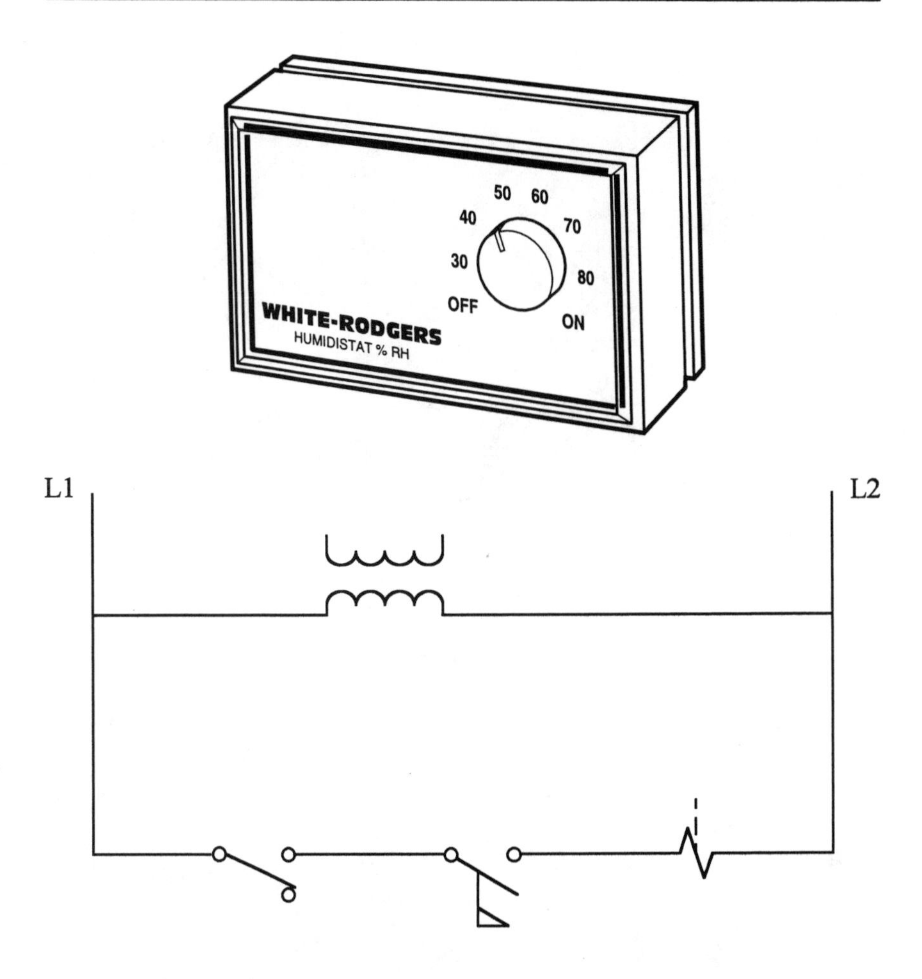

Figure 11-8. An example of a humidistat: (a) control unit, (b) circuit.

loss of line voltage, or low refrigerant. Short cycling should be avoided regardless of the cause because a certain amount of time is needed to equalize the pressure in a compressor when it is turned on. Short cycling also can cause excessive heat in the compressor due to the large current involved in each starting cycle.

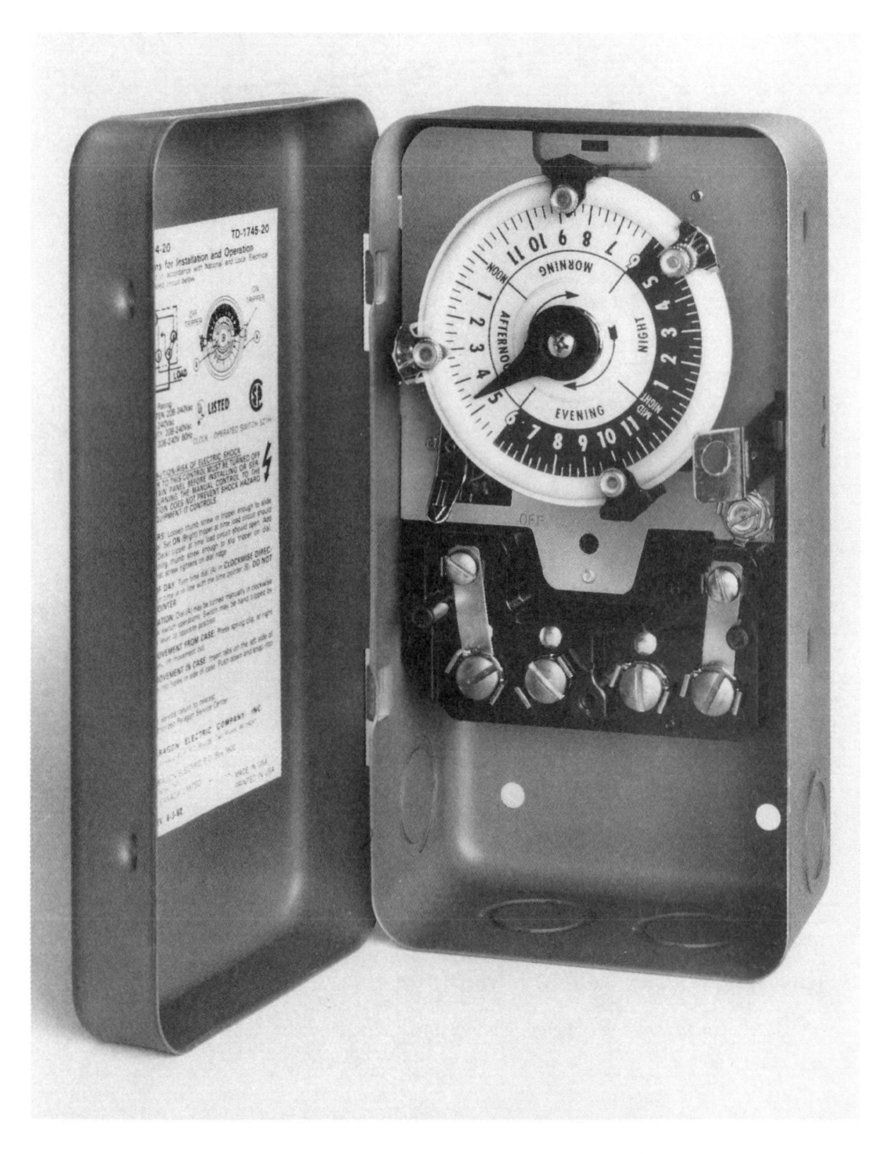

Figure 11-.9 An example of a programmable timer/thermostat. *(Courtesy, Paragon Controls)*

Figure 11-10 depicts an example of a short-cycle timer. The timer contains a set of contacts that open the line voltage if, for any reason, power to the compressor is interrupted. When power is reapplied the timer clock rotates through a set of gears and after a predetermined time operates a cam that closes a set of DPDT contacts. These contacts in series with other control contacts apply power to the compressor, and at the same time remove power from the timer motor.

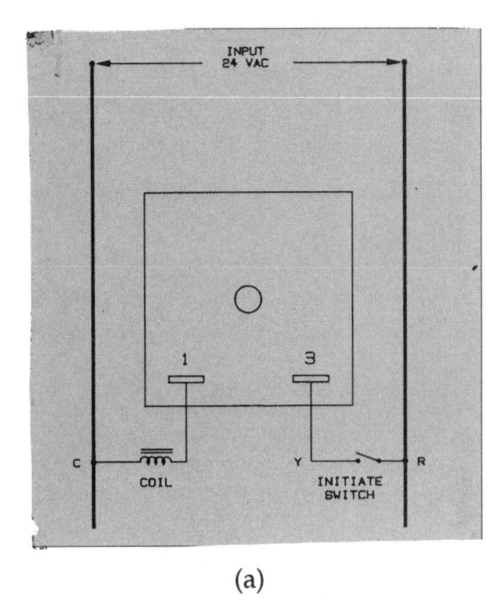

(a)

The defrost timer is used in automatic defrost or frost-free refrigerators and commercial freezers. The timer is operated by a synchronous clock motor that activates the defrost circuit each 24 hours. The clock

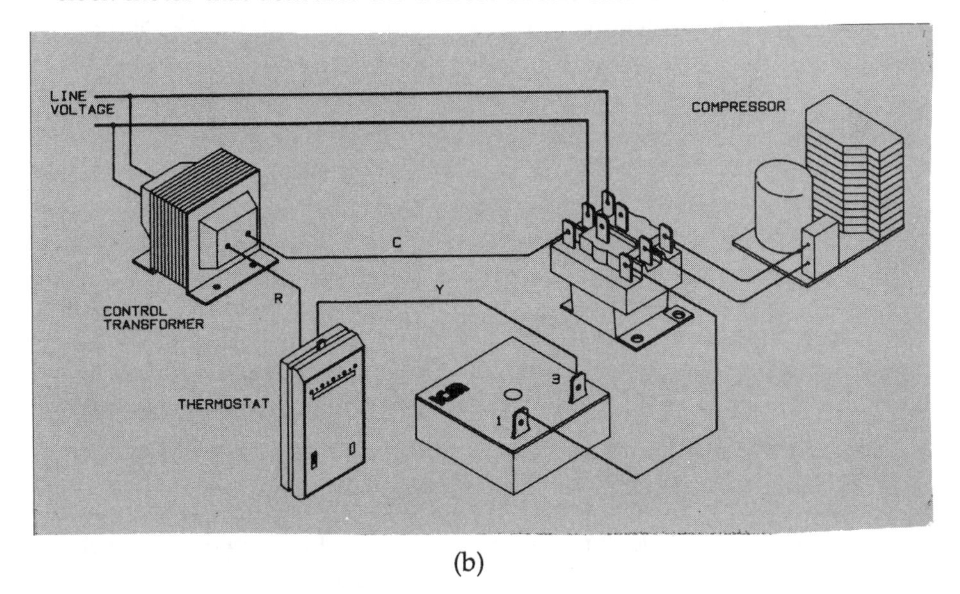

(b)

Figure 11-10. An example of a short-cycle timer: ladder diagram, (b) circuit application.

operates a cam that activates a switch that removes power from the compressor and applies power to the defroster heater. The timer allows this condition to last for a predetermined time after which the cam returns the switch to the normal position.

LIMIT SWITCHES

Limit switches are used in heating systems:

1. for safety to prevent excessive heat build up in the plenum,
2. to prevent cold air from being blown before the heat exchanger of the furnace has reached a predetermined temperature,
3. to keep the fan running until the heat exchanger has cooled to a predetermined temperature.

The most important limiting switch action is to prevent a fire from excessive heat in the plenum. This type of switch is usually an N.C. bimetal type snap switch that is connected in series with the 24-volt relay control of the main gas valve. In the event the plenum overheats, the thermal switch opens and removes the voltage to the main gas valve, shutting off the burner.

SOLENOID VALVES

A solenoid valve is a valve that uses the action of a solenoid coil to open or close the valve seat. Solenoid valves perform the same action as mechanical valves. However, they can be controlled automatically. Solenoid valves may be installed by sweat soldering or by a threaded input and output. An arrow is generally raised on the valve body to indicate the flow and assure correct installation.

A solenoid valve is shown in Figure 11-11. When in the energized position, current through the solenoid coil pulls the plunger into the coil and opens the valve seat allowing flow through the system. When power is removed from the coil, a spring closes the valve, stopping flow in the system.

THERMOPILE

A thermocouple can produce a voltage of approximately 30 millivolts DC. When more voltage is needed, thermocouples are added in series. These series connections are called thermopiles. An example of a thermopile is shown in Figure 11-12. These connections are packaged to develop 250, 500, and 750 millivolts.

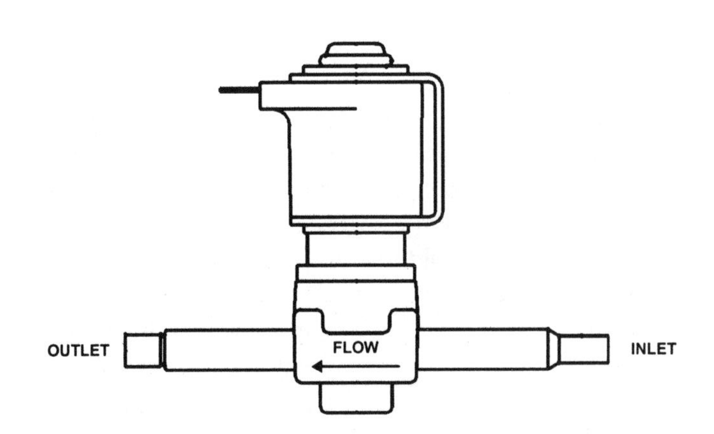

Figure 11-11. An example of a solenoid controlled valve.

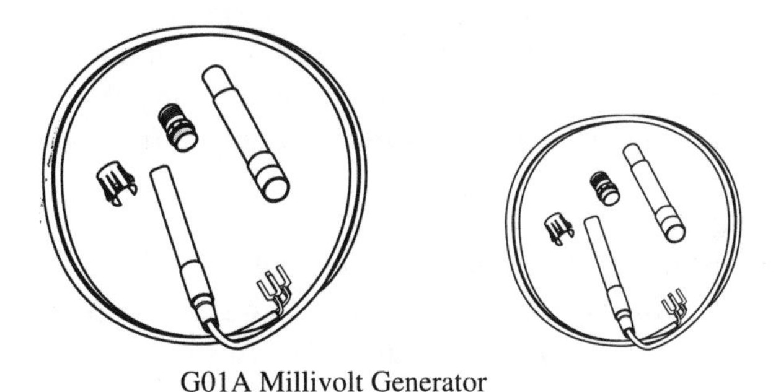

G01A Millivolt Generator

Figure 11-12. An example of a thermopile unit.

TROUBLESHOOTING

Thermostats

The operation of many thermostats can be observed by removing the cover of the device and varying the temperature selection dial. However, switch action is not assured unless an ohmmeter or voltmeter test is made on the output terminals.

Thermostats can be isolated by jumpering the connections at the output terminals at either the thermostat location or at the leads from the thermostat to the heating or air conditioning unit. Care must be taken when the thermostat is switching at high voltage.

Test procedures for computer-controlled thermostats are normally given with the manufacturer's installations or are on the inside of the unit cover.

Faulty thermostats are generally replaced.

Pressure Control Switches

Two common problems with pressure control switches are dirty contacts and leaks in the pressure input line or the bellows. The contacts can be tested with an ohmmeter or with a voltmeter. The voltmeter is placed across the contacts with power on. When the switch is in the open position, the meter should read circuit voltage. When the switch is in the closed position, the meter should read zero volts.

If it is impossible to make a voltage test across the contacts, a suspected pressure control switch should be replaced.

Sail Switches

A sail switch may be tested with an ohmmeter. The first test should be to assure that the sail switch has free movement. To test the switch contacts and sail action, the VOM is placed across the switch terminals and the sail lever manipulated by hand. The ohmmeter reading should swing from infinity to zero as the switch is changed from open to closed.

Humidistats

Humidistats are moisture-operated switches or relays and may be tested with a VOM, as any other switching. The problem of poor humidity control is often caused by the humidifier. This unit should be cleaned or replaced after testing the humidistat.

Timers

Timers are motor/cam operated switches or relays. These devices can be tested in a manner similar to any other switch by the use of an ohmmeter or voltmeter. The first test should be to determine that the timer motor is operating and that cam action is initiated at the selected time. After these tests the contacts may be tested for a continuity with a VOM.

Limit Switches

Limit switches are operated within a predetermined limit of motion or pressure. These switches may be tested with a VOM as any other switch. However, the limiting action that controls the switch must also be tested. It must be assured that the limit is reached before the switch can be tested, unless the limiting action can be positioned by hand.

Solenoid Valves

The switching action of a solenoid valve can sometimes be felt by placing a screw driver on the body of the valve as coil voltage is applied. Voltage may sometimes be measured at the coil terminals or at another location identified by the wiring diagram. As a last resort the coil leads may be opened and the correct voltage applied. Faulty solenoid valves must be replaced.

Thermopiles can be tested with a voltmeter if the output leads can be accessed. The proper heat must be applied to the unit.

SUMMARY

Safety controls and function controls are basically switches or relays. The uniqueness of each control is in the action or energy that controls the switching action. Although there is a commonality in the testing of these devices, each offers a different challenge.

QUESTIONS

1. What safety controls would be included in a newly installed air conditioning and cooling system for a large office building?
2. What is the purpose of a sail switch?
3. What is the purpose of platinum limit switches?
4. What is the advantage of anticipators in a thermostat?

Chapter 12

Motors

INTRODUCTION

Air conditioning and refrigeration systems make use of many AC motors. Motor action is also used in many devices that might not be thought of as motors. For example, watt hour meters, that are installed at the electrical service drop to most buildings, are a form of AC motors. All analog voltmeters, ammeters, and timing circuits operate on the motor principle. The computer mouse and even speakers for radios and televisions operate on the motor principle.

MOTOR ACTION

A motor is a device that changes electrical energy into mechanical motion. This is accomplished by a magnetic field rotating an *armature or rotor* on a shaft. The principle of motor action is shown in Figure 12-1. The armature consists of an electromagnet mounted on a shaft. The armature is free to turn between the poles of a permanent magnet called a *field magnetic or stator.* When the armature is in the position shown in the diagram, without a current through the winding, it does not turn. This is because it is attracted as much by the north pole as it is by the south pole of the field magnet. However, if we pass a current through the winding of the armature, an electromagnet is formed; the south pole of the armature will be attracted by the north pole of the field magnet, and the north pole of the armature will be attracted by the south pole of the field magnet. Therefore, the coil will turn one-quarter of a revolution.

When current is applied to the winding, the motor can turn in either direction. In a practical motor the magnetic field is distorted in some manner to cause the armature to turn in a certain direction.

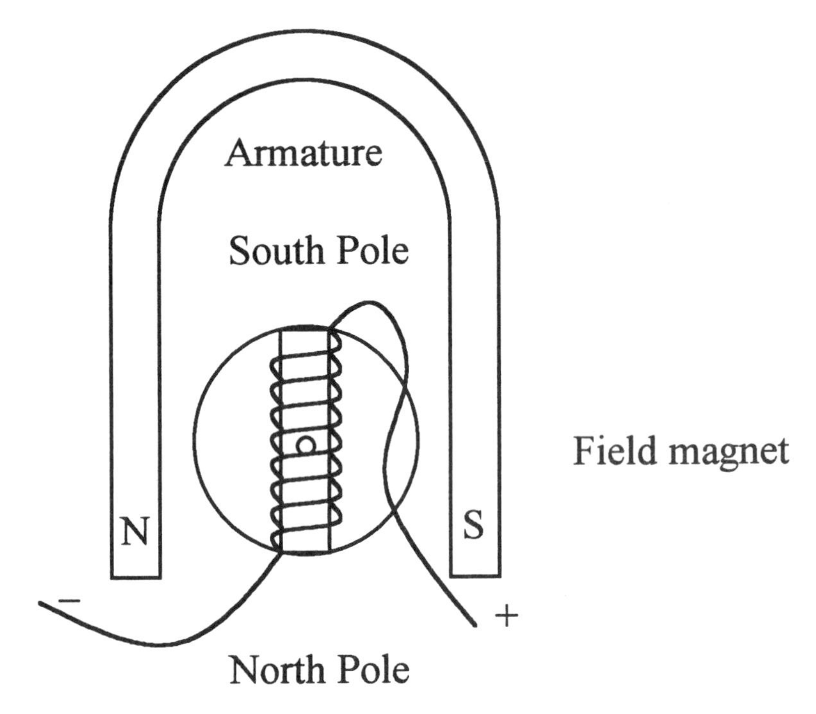

Figure 12-1. The armature rotates between the poles of the permanent magnet.

Suppose the armature in Figure 12-1 turns in a counterclockwise direction. The armature magnet will align *north-south* and *south-north* with the permanent magnet. The magnetic action would produce no more turning action because the sets of opposite magnetic poles are aligned with each other. However, because the armature has weight, and has been placed in motion, it continues to turn. At this point, the AC current reverses in the armature, causing the magnetic field in the armature winding to reverse. This action causes the armature to continue to rotate. In an AC motor the current in the armature reverses each half cycle of the voltage sine wave. A 60 Hz voltage would cause this simple motor to rotate at 3600 RPM.

Examining the operation of a DC motor may clarify the AC motor operation. In a DC motor the current reversals are made by an automatic switch on the armature, called a *commutator*. The arrangement of a communicator is shown in Figure 12-2. The commutator consists of two semicircles of copper connected to the armature windings, and mounted on the shaft that turns with the armature. Connections are made to the commutator through carbon or metal brushes that slide against the commutator as it rotates. This results in the current to the armature being reversed at the proper time to keep the armature turning.

SYNCHRONOUS MOTORS

AC motors operate much as DC motors. However, since an AC current reverses each half cycle, there is no need for a commutator to reverse current in the armature.

The *synchronous* AC motor is used where the motor must run at an exact speed. Examples of synchronous motors are clock and timing motors. Figure 12-3 illustrates the connection of a two-pole synchronous motor. The AC current flowing through the stator winding will cause the stator and armature to alternately attract and repel. The motor moves one-half a rotation for each cycle; therefore, the speed of the four-pole

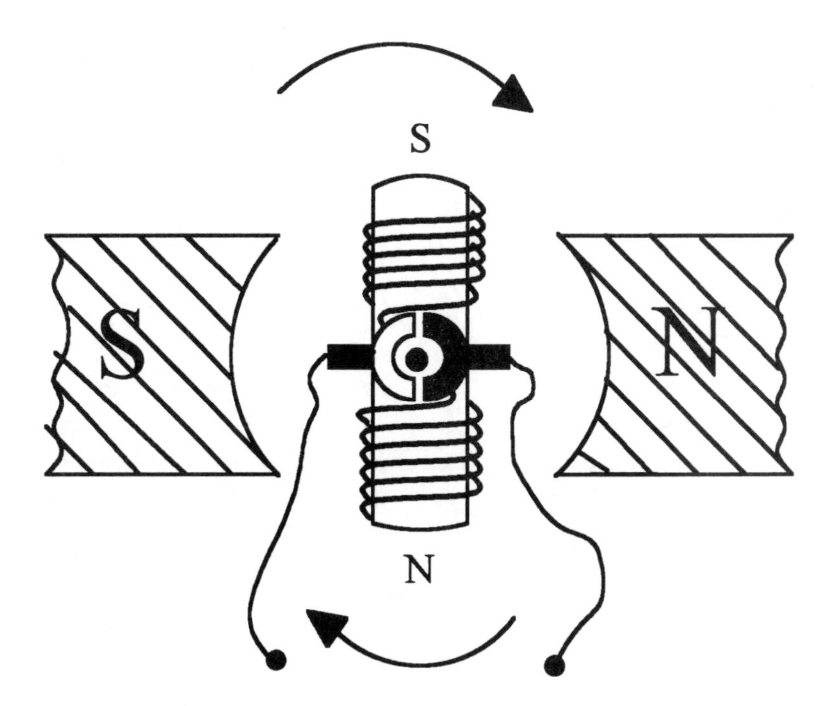

Figure 12-2. The commutator of a DC motor.

motor is 1,800 RPM. Synchronous motors use slip rings and brushes of carbon or metal to connect the AC current to the armature.

$$RPM = \frac{60\ Hz \times 60\ Sec}{2\ Pair\ Poles} = 1{,}800$$

INDUCTION MOTORS

Low-horsepower motors used for blower motors in residential environmental control devices are usually induction motors. An induction motor does not run at a constant speed, but is simple and comparatively economical to manufacture. The basic induction motor uses a pair of electromagnetic poles. The rotor of the induction motor is an assembly of copper bars and rings that look like a cage (Figure 12-4). Therefore,

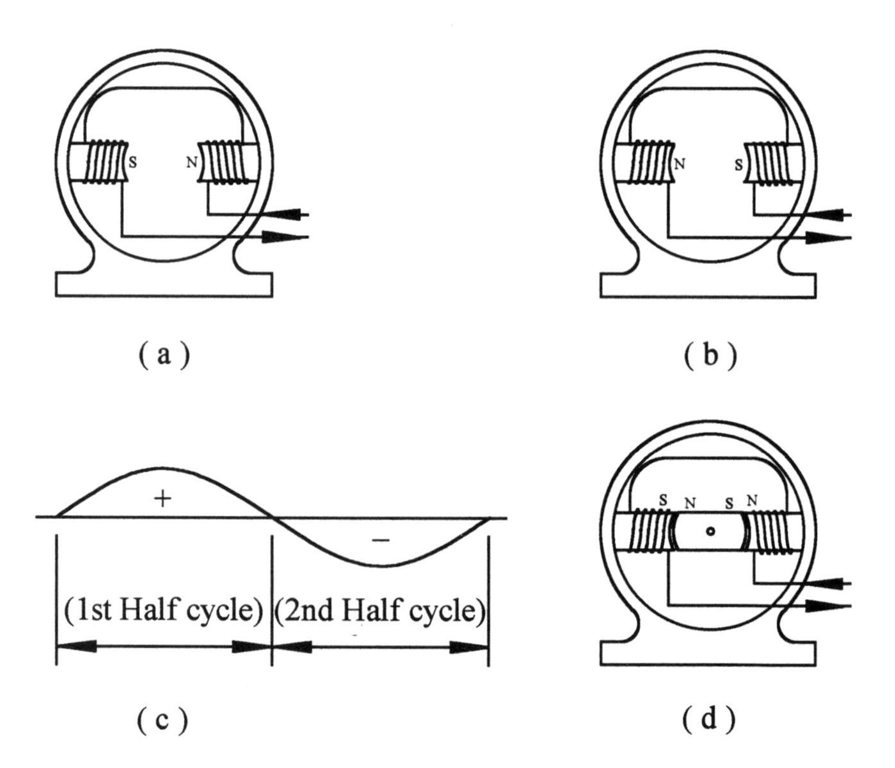

Figure 12-3. A two-pole synchronous motor.

the motor is called a *squirrel-cage* motor. Notice that the squirrel cage is basically a short-circuited winding that functions like the secondary of a transformer. There is no physical connection to either current or voltage.

An induction motor operates on the principle of transformer action. In other words, an AC current is induced in the rotor squirrel cage by the magnetic field from the stator poles. These induced currents also induce magnetic poles that are attracted and repelled by the stator poles. This action keeps the armature of the motor rotating. There is always a small amount of slip between the poles of the stator and the rotor so that the rotor is pulled and pushed.

Motors require some form of starting device to overcome inertia and assure that the motor will run in a certain direction. This is accomplished using a starting system. The shaded-pole motor in Figure 12-5 uses a metal ring around part of each stator field. When voltage is applied to the motor, a magnetic field is induced into the shaded area. This small magnetic field opposes the run field and distorts the main field to make the motor turn in a certain direction. The small shaded magnetic field has little effect on its operation when the motor reaches run speed.

SPLIT-PHASE MOTORS

Split-phase/split-phase motors are used in applications of less than 1/4 horsepower that require relatively low starting torque. These motors are used in applications where this is not a restriction, such as refrigerator compressors and fans. In a split-phase motor the current is split be-

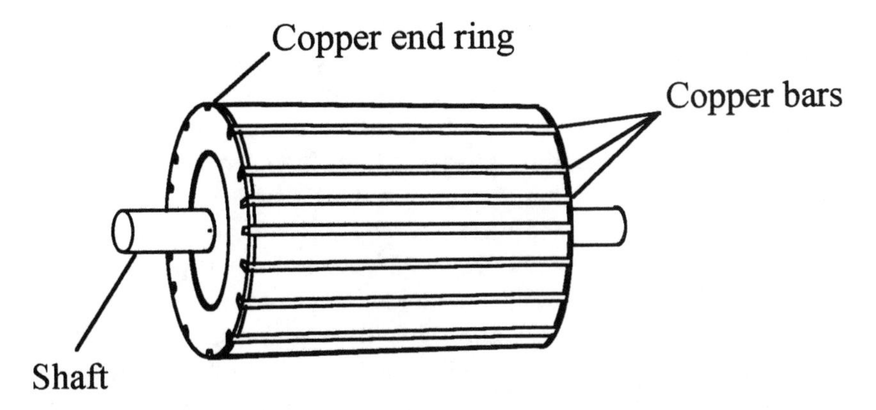

Figure 12-4. The construction of a squirrel-cage rotor.

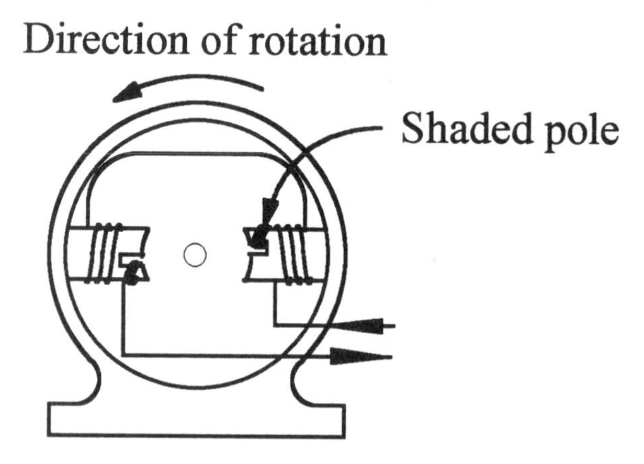

Figure 12-5. Example of a shaded pole motor armature and field.

tween two windings which are arranged to develop current that is out of phase.

There are four types of split-phase motors: the resistance start inductance run (ISIR), the capacitor-start inductance run (CSIR), the permanent capacitor (PSC), and capacitor-start capacitor-run (CSCR).

Figure 12-6 shows the winding arrangements for a split-phase motor, and the phase relation of the resulting currents. When current is applied to the motor, both the starter winding and the run winding produce currents that are 90° out of phase. The starter winding gives the added energy to overcome inertia. When the motor reaches approximately 45% of running speed the starter winding is removed by way of a centrifugal switch (Figure 12-7). The switch is composed of a spring-loaded arrangement that closes the contacts to the starter winding when it is at rest, and opens the contacts when the armature reaches running speed. The contacts of this switch often become corroded, preventing current from flowing in the starter winding. The symptoms are a motor that will not start and often hums. For a short-term emergency solution, the contacts can be cleaned with fine emery cloth or a burnishing tool. However, the mechanism should be replaced as soon as possible.

Another starting arrangement for a low-power split-phase motor is the current relay shown in Figure 12-8. The high starting current in the run winding closes the starting relay contacts. When the motor reaches full speed the current in the run winding decreases and allows the start relay to open.

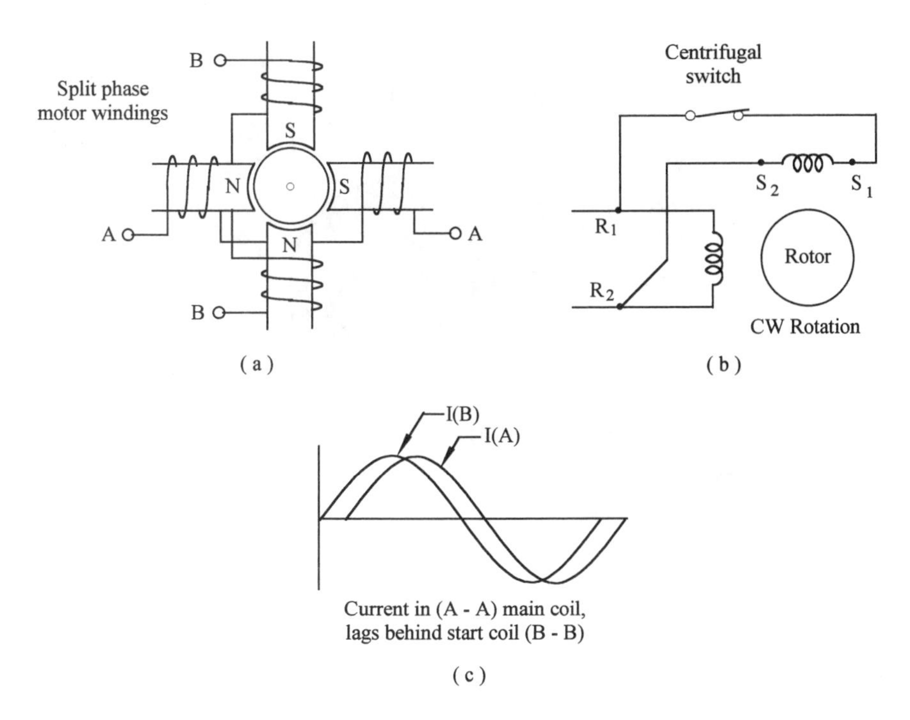

Figure 12-6 The winding currents and magnetic fields are 90° out of phase.

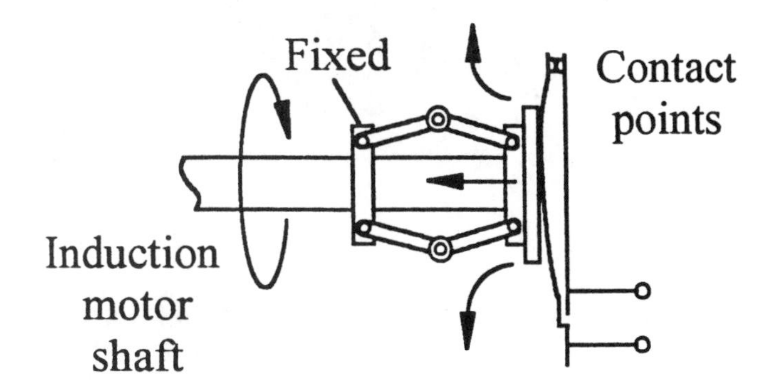

Figure 12-7. Centrifugal switch starting devices.

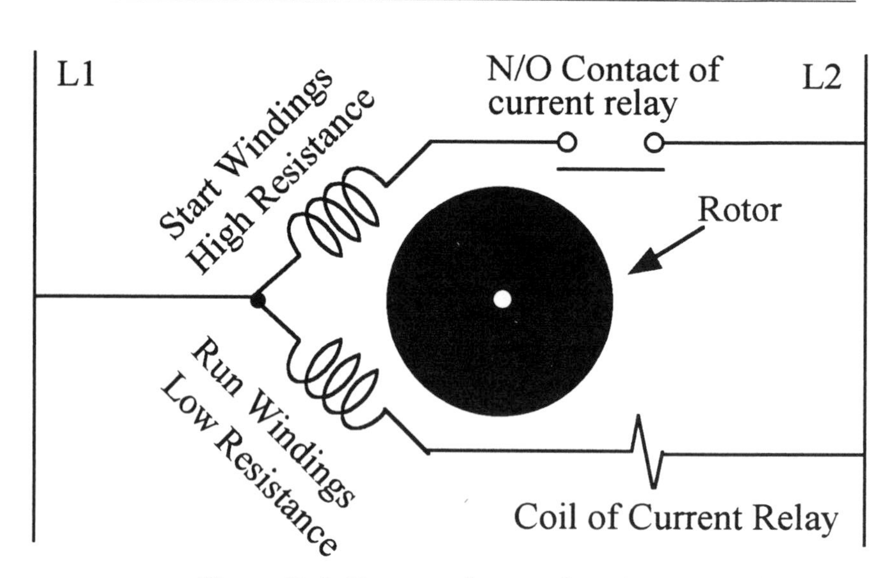

Figure 12-8. Current relay starting circuit.

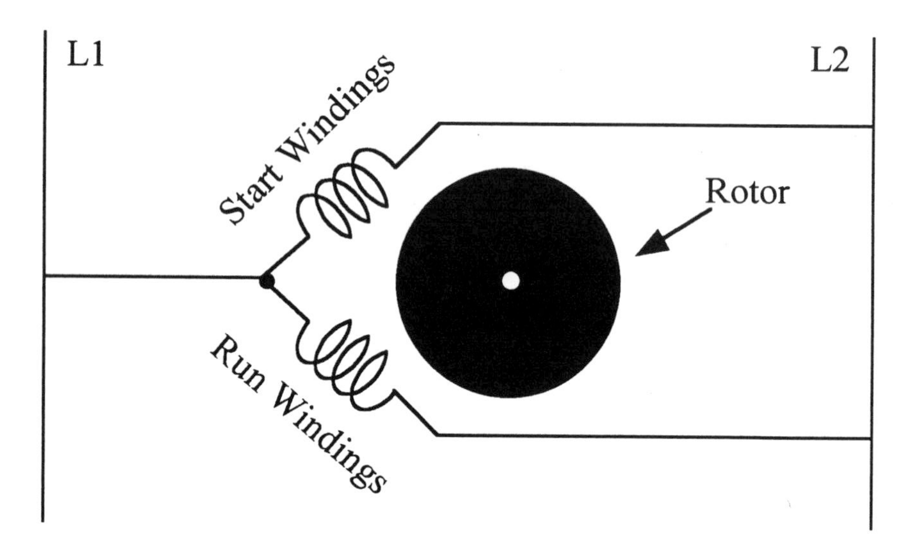

Figure 12-9. Inductor-start inductor-run wiring diagram.

Figure 12-9 depicts the wiring diagram of the inductor-start inductor-run motor circuit. Both windings remain in the circuit during both start and run.

CAPACITOR-START MOTORS

The capacitor-start motors use a starting capacitor, and a centrifugal switch or a current relay connected in the starting winding. The addition of a capacitor in the starting winding increases the winding current and phase shift for increased torque. The starting winding is sometimes referred to as an auxiliary winding. Figure 12-10 is a schematic diagram of a capacitor-start motor. These motors are available in 1/4 to 3/8 horsepower ratings, and find wide use in environmental control systems.

The start relay in Figure 12-10 is a current relay that operates by sensing the magnitude of the current in the circuit. The current relay is

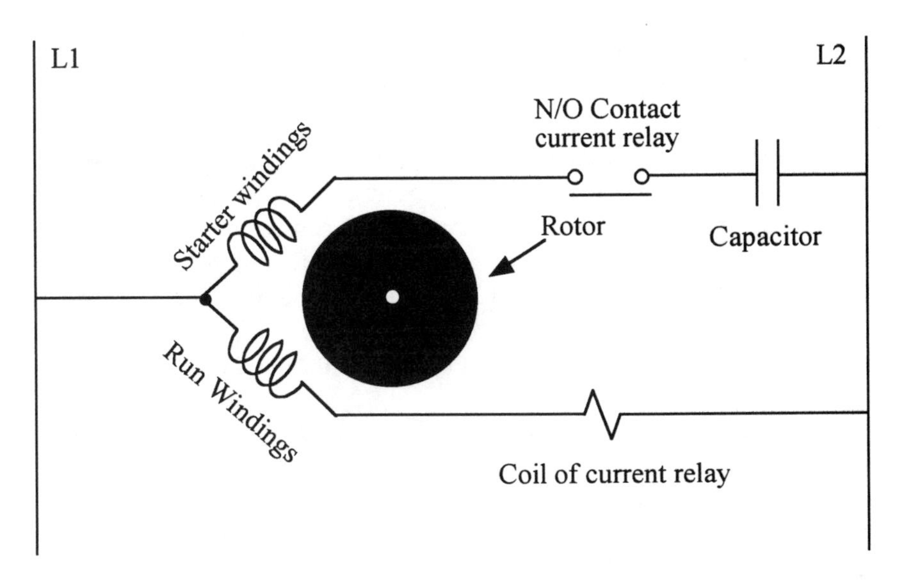

Figure 12-10 Schematic of a capacitor-start motor.

comprised of a coil of a few turns of heavy wire and a set of normally open contacts. The coil is connected in series with the start winding. When power is applied to the motor the start contacts are open and no power is applied to the start winding and the motor fails to start. Since the armature is stationary, a current of approximately three times normal flows through the run winding and the start relay coil. This results in a strong magnetic field that closes the start contacts of the relay. As the contacts close, current flows through the starting winding causing enough torque to start the motor. When the motor reaches approximately 75% of full speed, the counter-electromotive force produced by the turning action causes the current in the run winding to decrease to a level that no longer holds the start relay closed. The start relay is disconnected and the motor runs normally. If for any reason the motor speed should decrease, the potential will again engage the start winding. Procedure to test a current start relay will be presented in Chapter 13.

CAPACITOR-START CAPACITOR-RUN (CSCR) MOTORS

The capacitor-start capacitor-run (CSCR) motor depicted in the schematic in Figure 12-11 is essentially a CSIR motor with a capacitor that remains in series with the start winding after the start capacitor is opened.

As with the CSIR motor, the starting mechanism removes capacitor C1 when the motor reaches speed. Capacitor C1 is an electrolytic starting capacitor and capacitor C2 is an oil-filled run capacitor. A potential relay is used to disconnect the start capacitor. Examples of these two types of capacitors are shown in Figure 12-12. The direction of rotation of the CSCR motors can be changed by reversing the connections of either the starting or the running windings, when the ends of the windings are exposed.

The starting torque is achieved by starting capacitor C1 and the run capacitor C2 in parallel and in series with the start winding. When the motor starts, the run winding (RW), the start winding (SW), the run capacitor (RC), and the start capacitor (SC) are all connected. The parallel combination of the RC and SC in series with the start winding causes a large current in the start winding, resulting in a large starting torque. When the motor reaches approximately 75% of speed the induced voltage in the run winding energizes the start relay, the NC contacts open

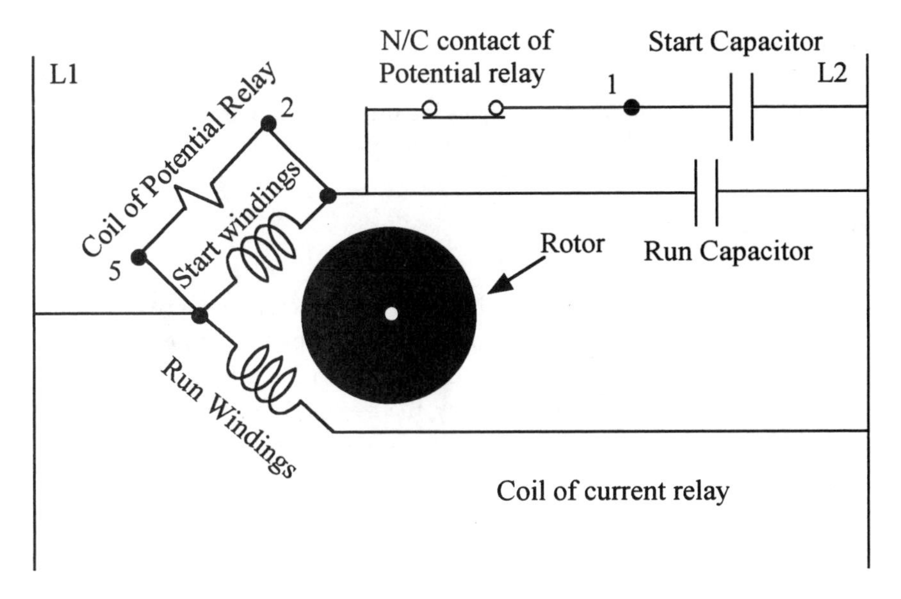

Figure 12-11 A schematic of a capacitor start capacitor run (CSCR) connection.

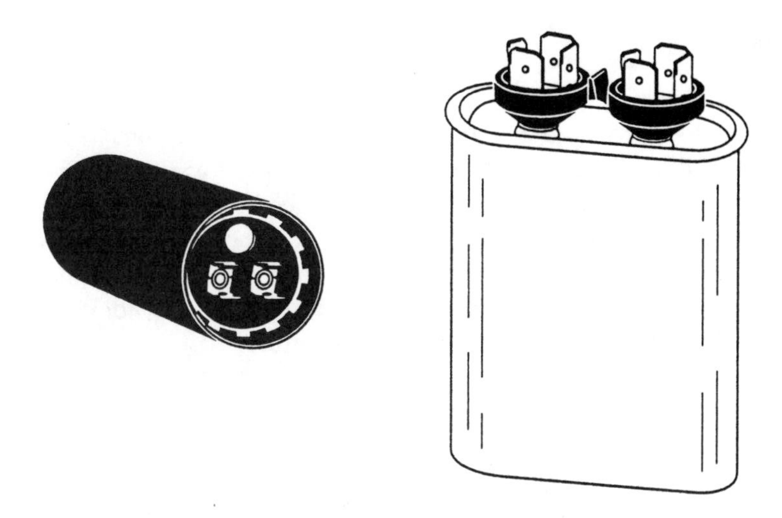

Figure 12-12 Motor capacitors: (a) an electrolytic start capacitor, (b) an oil-filled run capacitor.

and remove the start capacitor. This type of relay cannot be used in a motor where the start winding is disconnected from the circuit. That type of motor circuit would de-energize the relay and close the NC contacts, and again apply starting current to the motor.

DUAL-SPEED MOTORS

Dual-speed motors used in heat/cooling circuits have two or four sets of stator poles. At normal speed the two poles are connected in the circuit. At low speed the motor is connected to have four poles in the stator. Motor speed is determined by the demand, and is usually controlled by a thermostat. When demand is low, the system operates at a lower speed, and is more efficient, and is cost effective. Motor speed is calculated as follows:

$$RPM = \frac{2f \times 60}{NP}$$

Where f = frequency in Hertz (cycles per second), 60 converts cycles per second to minutes, and NP = number of poles.

For example: A motor operates on 240-volts at 60 Hertz, and has 6 poles: the motor speed is:

$$RPM = \frac{2 \times 60 \times 60}{3} = 1{,}800$$

Multi-speed fan motors make use of multiple windings in series as shown in the schematic in Figure 12-13. The speed control switch adds or removes winding. To obtain a higher speed, the control switch removes a winding causing greater current and increased speed. Newer multi-speed fan motors utilize solid-state motor controllers. Solid-state controllers make use of semiconductor circuits that control devices such as the SCR, DIAC and TRIAC to drive the motor.

MOTOR SLIPPAGE

A motor has the tendency to lag behind the magnetic field produced by the AC voltage. This slippage is the amount that a motor's speed is slower than the calculated value. Under no load, slippage is

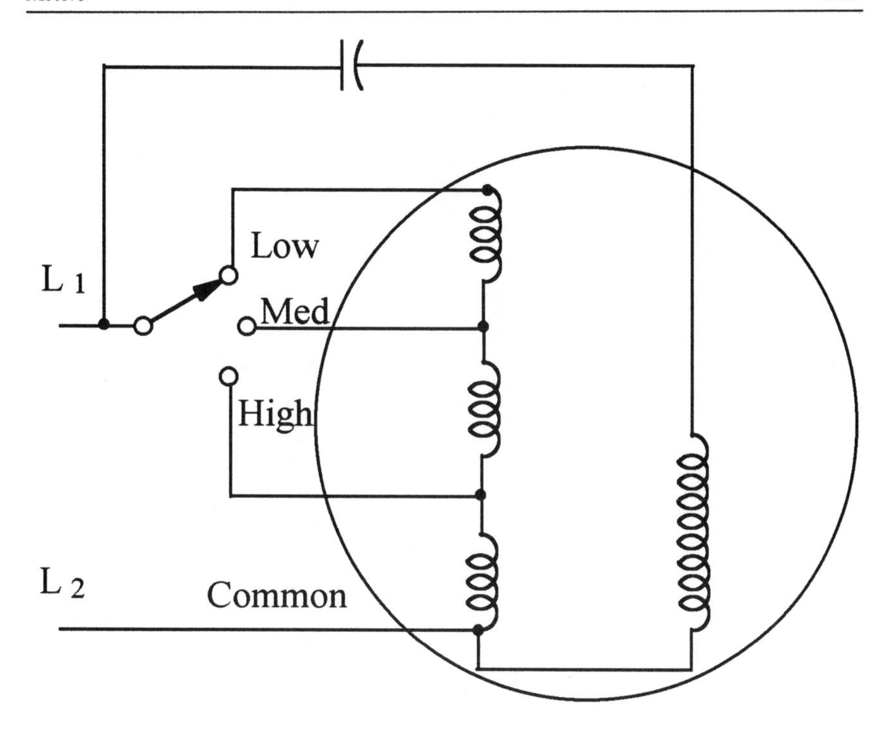

Figure 12-13. An example of a dual-speed fan circuit.

small. However, as the load is increased, slippage increases and the motor slows. Under normal load we may expect a slippage of 3 to 5 percent. For example, under normal load, a motor that is calculated to operate at 1,800 rpm would have a speed of approximately:

$$RPM = 1{,}800 \times 0.95 = 1{,}710$$

The true operating horsepower of a motor depends to a great extent upon the operating conditions of the system. The power factor of the line voltage, the efficiency of the motor, and the power consumed must be considered. The true horsepower of a single-phase motor is formulated.

$$HP = \frac{E \times I \times \%Eff \times PF}{746}$$

The true power for a three-phase motor is formulated:

$$HP = \frac{E \times I \times \%Eff \times 1.732}{746}$$

THREE-PHASE MOTORS

Three-phase motors have three separate stator windings of the same number of turns and resistance. The windings require voltage and current from three phases of electricity 120 degrees apart. Unlike single-phase motors, three-phase motors do not require start windings, starting relays, or starting capacitors because the three currents in the windings are 120 degrees apart. The three phases of current produce a greater start and run torque than single-phase.

The motor windings can be connected in either the delta (Δ) or wye (Y), as were three-phase transformers. The wye connection shown in Figure 12-14 places the winding in each phase in series with the other windings of that phase. The delta connection, shown in Figure 12-14(b), places the each winding across a phase voltage and increases winding current. This arrangement decreases the winding resistance to approximately 1/4 that of the wye connection.

Wye-connected motors are commonly used because they draw less starting current than delta connections. Some three-phase motors use a wye connection for starting and switch to a delta for running. This arrangement results in lower starting current and increased running power.

The synchronous speed of a three-phase motor is determined by the number of poles and the line frequency. The speed of a motor operated from a 60 Hz line is:

2 Poles = 3600 RPM
4 Poles = 1800 RPM
6 Poles = 1200 RPM
Poles = 900 RPM

Three-phase motors are used on commercial and industrial systems. Many three-phase motors are designed to operate on either 220 volts or 440 volts. This is accomplished by utilizing multiple windings that may be connected either in parallel or in series. The three-phase motor in Figure 12-15 (a) is connected for 220-volt operation. The stator

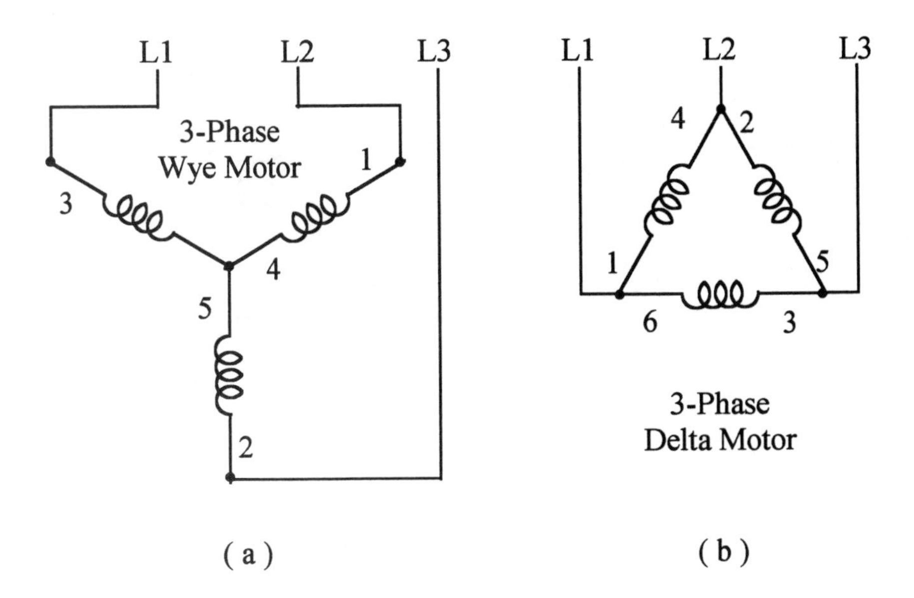

Figure 12-14. Three-phase motor connections: (a) wye connection, (b) delta connection.

windings are rearranged into a series connection (higher resistance) to operate on 440 volts (Figure 12-15(b).) The motor instruction label will indicate the connection changes that must be made to convert a motor from 220-volt to 440-volt operation. *These instructions must be followed exactly.*

Some three-phase motors have six accessible leads that can be connected in either the delta or wye configuration. Typically the leads that are accessible in the motor terminal box are labeled 1 through 6. Figure 12-15(a) shows the configuration for a wye. Power is applied to leads 1, 2, and 3. The resistance between 1-4, 2-5, and 3-6 will be the same.

Figure 12-5(b) shows the hookup of the six windings to form a delta connection. Power is applied to 1-6, 2-4, and 3-5. Although the delta connection requires more starting current than the wye connection, it has more running power.

Problems to air conditioning, refrigeration, and heating systems are often caused by motor failure. A detailed study of motor failures and motor repair will be conducted in Chapter 13.

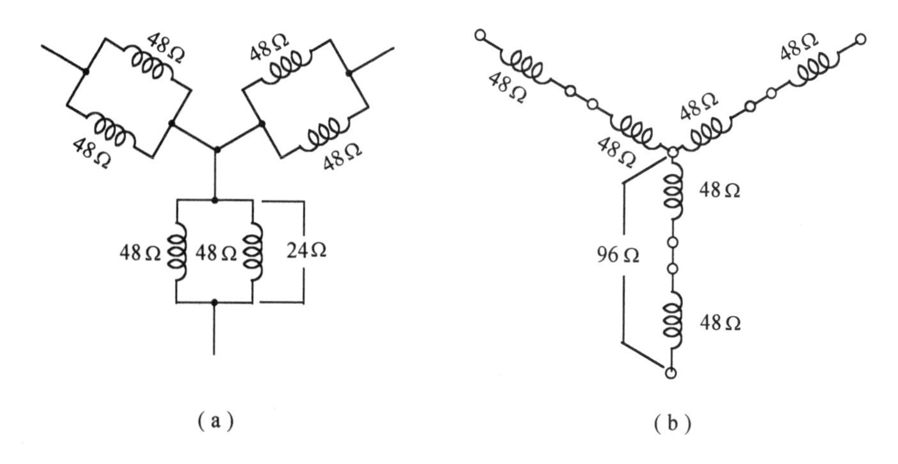

Figure 12-15. Conversion of a three-phase motor from 220 volts to 440 volts.

SUMMARY

1. Motors operate on the magnetic principle of attraction and repulsion.

2. Motors that are taken out of service should be dissembled, cleaned and bench tested.

3. The ohmmeter is used for continuity tests and for measuring winding resistance.

4. The megger develops a high voltage and measures leakage resistance in the millions of ohms.

Chapter 13

Motor Failure and Repair

INTRODUCTION

Motors can be very reliable. However, many have constant usage, and operate in an environment which may include excessive dust and heat. Many motors that you encounter will have been operating for thousands of hours. Motors are among the most-often-failed components in environmental control systems. However, when encountering a suspected motor problem, consider all other possible causes. Always examine the problem carefully, and look for the simple causes, such as loose or broken wires, poor connections, a faulty starting relay, faulty thermocouple or a defective thermostat.

Motor failures may be classified in one of five categories: *overheating, not starting, intermittent operation, loss of power, and noisy operation.* Causes of these failures and their corrections are reviewed.

OVERHEATING

Overheating of a motor is caused by increased current in the windings resulting in increased power dissipation and excessive heat. There are many problems that cause overheating. The technician's task is to determine the source of the problem and make the correction. Possible sources of overheating:

- Dirt and dust collecting in the motor or starting device, preventing the proper flow of cooling air through the motor housing.

- Overload of the motor by excessive duty cycle demand, or a motor that has insufficient horsepower for the load.

- High or low voltage. High voltage results in increased current and power loss in the motor. Low voltage results in decreased horsepower, causing an overload to the motor.

- Short cycling on and off. The starting cycle of a motor requires two to three times the power of the running cycle.

- Loss of one phase of a three-phase system. This causes excessive current in the remaining windings of the motor, resulting in increased power dissipation in the motor.

- Restricted ventilation, preventing proper cooling, leading to excessive heat and reduced motor life.

- Voltage imbalance in a three-phase system can cause excessive loss of power and overheating.

- Shorted turns in the motor stator causing excessive motor current.

- Failure of the starting capacitor, preventing the motor from operating.

- Failure of the run capacitor, resulting in excessive current and heat.

MECHANICAL FAILURE

Mechanical failures are usually easy to detect. However, the cause of the failure should be determined and corrected. Some causes of mechanical failure are:

- Moisture or water induced into the motor causing bearing failure, overload, or internal shorts.

- Surge voltage causing internal shorts in the stator windings, resulting in overheating or loss of operation.

- Bearing failure due to dirt, moisture, or lack of lubrication. Bearing failure can usually be detected by sound. However, bearing failure may only result in an effective load increase.

- Vibration may cause bearing failure.
- Worn drive belts or a bent shaft may cause bearing failure.
- Internal starting device contacts may become corroded and effectively open the starting winding. Each time the motor starts the electromagnetic induction of the starting current causes a large arc across the contacts. Failure is caused by the number of times that the motor cycles on and off, and may occur in a relatively new motor that has short cycling.
- Failure of the starting capacitor, preventing the motor from operating, may be considered as a mechanical failure.

TESTING MOTORS

The main concern when testing motors and motor circuits should be safety. We will consider a number of types of motor failure and the proper repair of each. Mechanical problems are usually obvious to the technician. However, the repair may be difficult. Several questions should be answered before determining that the motor is at fault.

1. Is there power to the motor?
2. Is the starting capacitor faulty?
3. Are the overload safety devices operating correctly?
4. Are the safety interlocks closed?
5. Are there loose connections?

NOISY, FROZEN, OR VIBRATING BEARINGS

Motor bearings are normally easily replaced except in sealed compressor units. Before removing the motor to the workbench, disconnect the main power to the unit. With the motor on the bench, mark the orientation between each end bell and the stator housing so that they may be correctly aligned later. Remove the end bells from the motor. The rotor can be attached to one of the bell housings by the bearings (Figure 13-1). At this point all parts should be cleaned of dust and dirt with a brush or air. When air is used the technician should wear protective goggles and mask. Remove the bearing and obtain a replacement. Replacements may be obtained from the manufacturer, a parts warehouse, or a bearing supply house. A bearing dealer is usually a faster and less expensive source. Most motor bearings are the self-lubricating type. Finally, reassemble the motor and test its operation, following all electrical safety procedures.

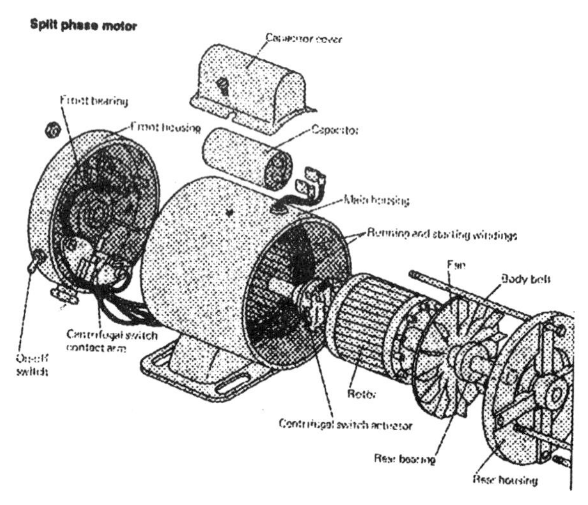

Figure 13-1. A disassembled motor.

STARTING MECHANISM

The centrifugal single-phase starting mechanism, Figure 13-2, is the most common cause of trouble with motors using that device. High resistance of the contacts limits the starter current, preventing the motor from starting. In rare cases the contacts may weld together and keep the starter winding engaged, causing the motor to overheat. The contacts can be polished with very fine sandpaper or a burnishing tool.

TESTING MOTOR WINDINGS

Motor windings are best tested with an ohmmeter. The winding resistance can be measured with the ohmmeter. The greater the winding wire size, the less the resistance.

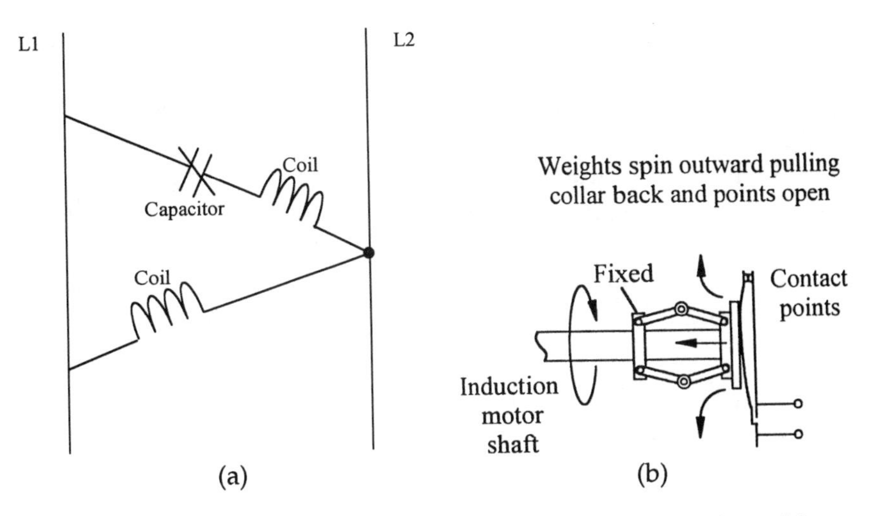

Figure 13-2 Internal starting mechanism: (a) schematic view, (b) example device.

An open winding will read infinity (Figure 13-3). When making the measurements, do not place your fingers on the ohmmeter tips as this will parallel your body resistance and give a faulty reading. A grounded stator winding can be located with the ohmmeter as shown in Figure 13-4. Any reading other than infinity between any stator winding and the

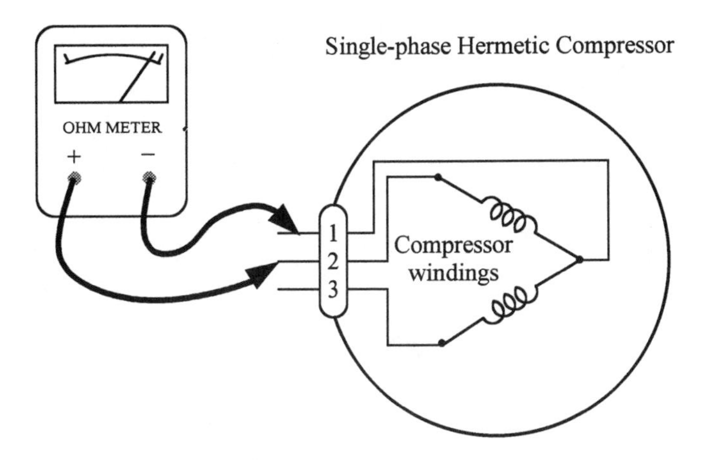

Figure 13-3. Testing for an open winding.

stator indicates an internal short. The short may be located by careful visual inspection for insulation wear or break. The insulation can be repaired by the application of liquid plastic or liquid rubber over the break.

High resistance leakage can occur between motor windings or between windings and ground. The leakage may occur only when there is a high voltage present. In that case, an ohmmeter will not detect the fault, and a high-voltage ohmmeter must be used. This device called a *megger* produces 500 volts and reads resistance in the millions of ohms.

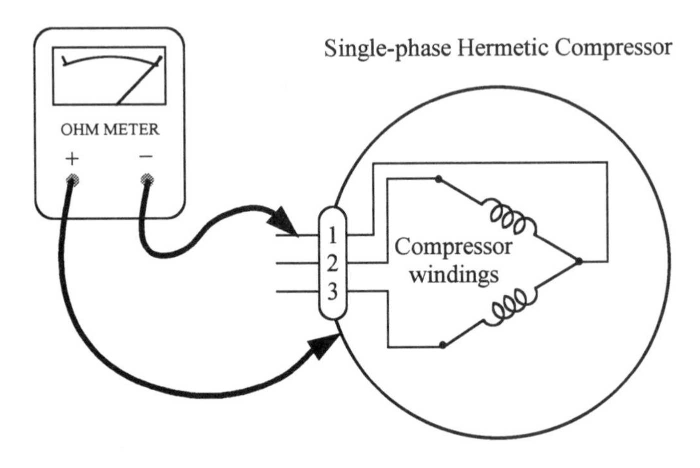

Figure 13-4. Using an ohmmeter to test for a grounded winding.

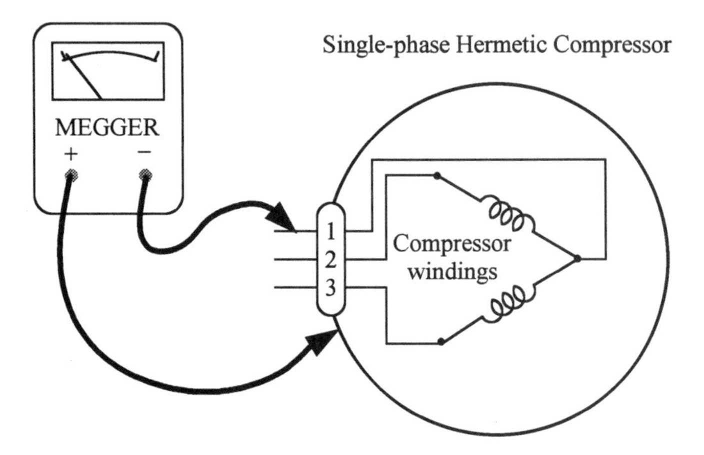

Figure 13-5. Using a megger to test for high leakage resistance.

megger is used exactly as any other ohmmeter. However, it must never be used on low-voltage semiconductor circuits, because the voltage will destroy them. Figure 13-5 is an example of a high-voltage megger.

Most domestic refrigerators and freezers use hermetic type compressors where the motor and starting device is completely enclosed. The three connections, Figure 13-6, are the only access to troubleshooting the motor.

WOUND ROTOR INDUCTION MOTORS

The wound rotor motor uses slip rings to connect the applied voltage to rotor windings that develop the rotor magnetic field. In addition to inspecting the starting device, the brushes should be checked for wear and weak brush springs before assuming the problem to be the motor windings. Most large motors of this type use a stepping device to limit the current in the winding. As the motor gains speed, the rotor winding offers greater opposition to current, allowing the starting resistance to be

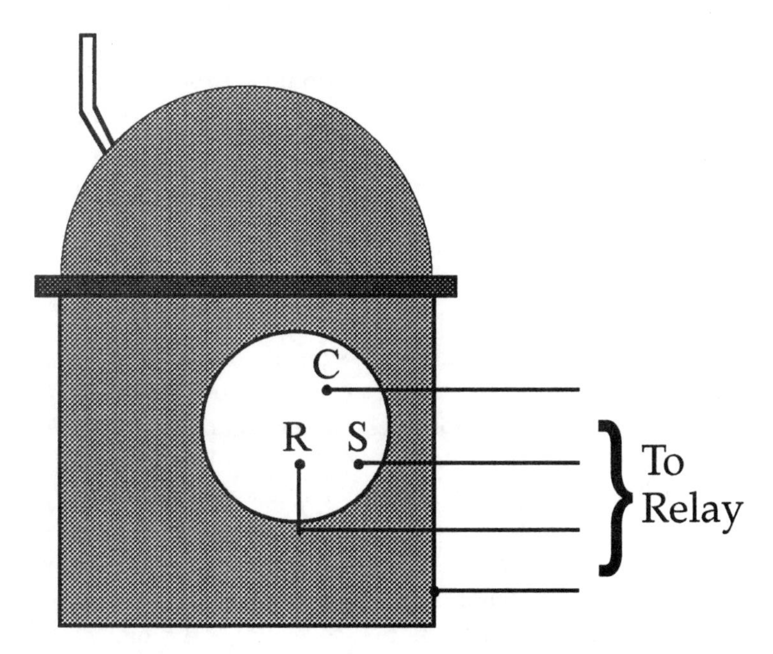

Figure 13-6. Terminal connections of a split-phasehermetic type compressor.

reduced. When the motor reaches full speed all the starter resistance is removed. If a wound rotor induction motor does will not reach full speed, the problem is usually the starter step device.

Testing a wound rotor induction motor requires that both the stator and the rotor windings be tested for shorts, opens, and grounds. The winding ohmmeter reading should be very low between the slip rings. The reading between each of the slip rings and ground should be infinity.

TESTING MOTOR CAPACITORS

The theory of capacitors and motors has been covered in earlier chapters. We will make only a brief review of these topics here. Capacitors have the ability to store electrons, and shift the phase of current and voltage. This shifting of the phases of the currents aids the motor in starting and in some cases both starting and running.

Split-phase motors have starting and running windings. The starting windings are used to overcome the inertia of the motor and load. The starting winding is comprised of many turns of fine wire and has a much higher resistance than the running winding. This higher winding resistance causes the currents in the starting and running winding to have approximately a 60-degree phase difference.

The starting winding is removed from the circuit as the motor approaches its running speed. This is accomplished by the centrifugal switch. The spinning action forces the contact points open as the motor reaches running speed when the added torque is not needed.

A start capacitor added to the circuit causes additional phase shift of the starting current, and increases the starting torque. Starting toque is 300 to 500 percent as compared to that of an inductance-start motor. As with the inductance-start motor the starter winding is removed at approximately 3/4 full speed.

The *capacitor-start capacitor-run* motor utilizes both a starting capacitor and a run capacitor. During the starting cycle both capacitors are in series with the start winding. When the centrifugal switch opens, at 3/4 speed, the starting capacitor is removed. However, the run capacitor remains in series with the run winding, correcting the motor's power factor.

The permanent-split capacitor motor (PSC) is a split-phase motor with a run capacitor winding. No starting switch is used and the capacitor and starting winding are in the circuit at all times.

Start capacitors are rated in micro-farads and working voltage. Typical ratings are 120 volts to 330 volts and 21 to 1200 micro farads. Start capacitors are designed for intermittent operation. Starting capacitors are designed for low cycling and short duration starts. Typically, a maximum of 3 seconds on and 20 starts per hour. One terminal of a starting capacitor is connected to the case. A failed starting capacitor must be replaced with one of the proper size and voltage rating.

Run capacitors are electrolytic type capacitors that are designed for continuous duty. They normally have less capacitance than starting capacitors.

Most capacitor failures are the result of over cycling, over heating, or age. These failures exhibit themselves in leakage of the electrolytic oil, reduced capacitance, open capacitance and shorted capacitance. Any capacitor that shows physical damage or is leaking oil should be replaced. The replacement must be of the same value and voltage rating. However, the voltage rating of the replacement can be higher than the original.

An ohmmeter can be used to test for most capacitor failures. To perform a test on a capacitor the ohmmeter should be on the highest scale. The meter should read a low value and slowly increase to infinity, when the leads are placed across the terminals of a disconnected capacitor. A shorted or leaky capacitor will cause a low reading. An open capacitor will result in the meter instantly reading infinity.

Table 13-1 is a flow chart that can be used to locate motor troubles caused by capacitors. The chart is used by starting at the top and following the flow as you answer the questions. Power must be disconnected before making any resistance tests on the system.

- An internal short between windings is usually impossible to find with an ohmmeter. However, a careful comparison of winding resistance measured with a DVM may reveal the information.

- The DVM or VOM are standard equipment for the technician.

- The most common method of testing a motor in the field is by using an ohmmeter.

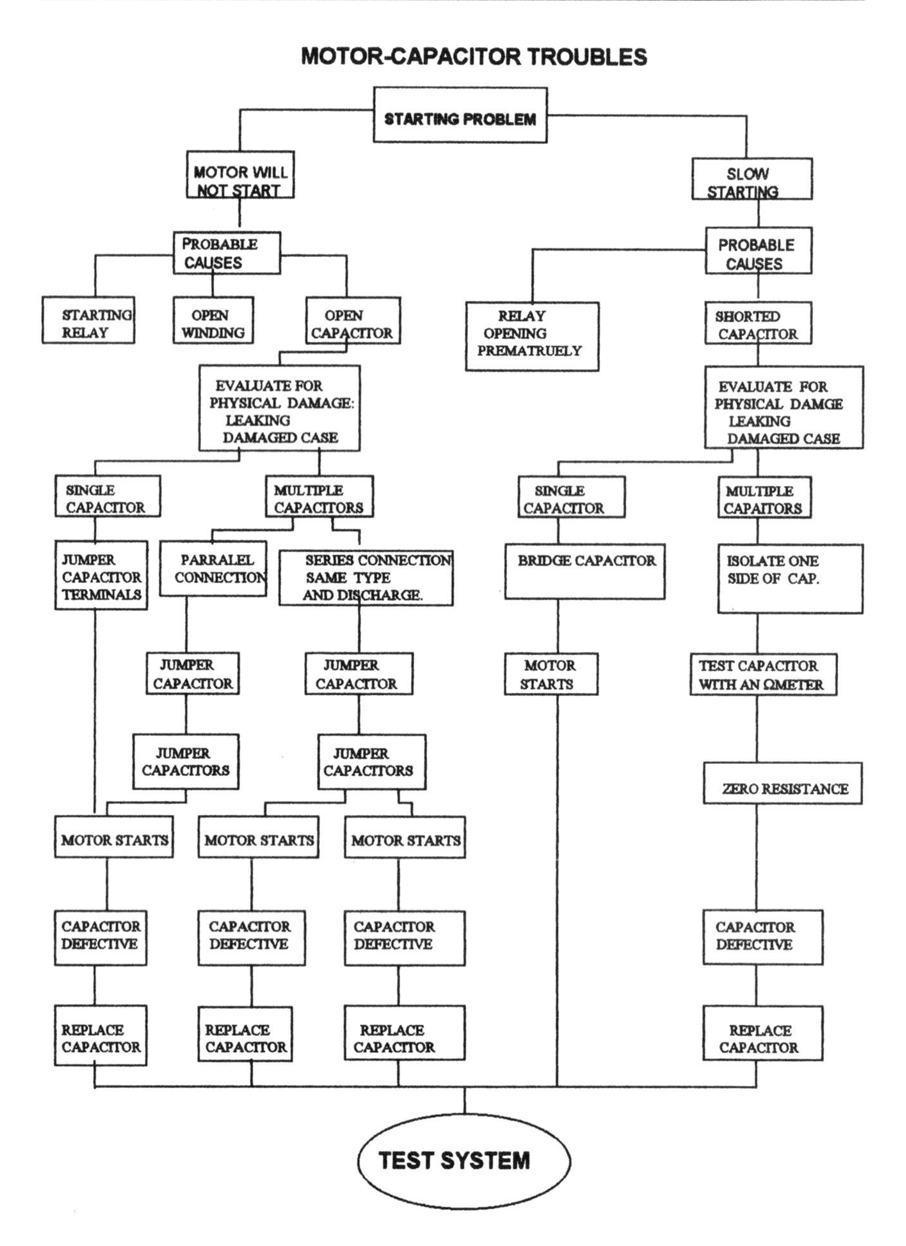

Table 13-1. A flow chart to locate motor capacitor failures.

- Power must be removed from the circuit when using an ohmmeter.
- One of the windings or components being measured with the ohmmeter must be isolated.
- A motor's winding resistance is usually measured on the low ohms scale.
- If no reading is obtained on the low scale, use a higher scale.
- Your fingers should not touch the ends of the probes when measuring resistance.
- Clamp-on ammeters are connected around one line of a supply or winding.
- The voltmeter is always connected across a component or from the line to ground.

SUMMARY

1. Motors operate on the magnetic principle of attraction and repulsion.
2. Motors that are taken out of service should be disassembled, cleaned and bench tested.
3. The ohmmeter is used for continuity tests and for measuring winding resistance.
4. The megger develops a high-voltage and measures leakage resistance in the millions of ohms.

QUESTIONS

1. What is the basic principle of magnetism on which motors operate?
2. What is the purpose of brushes on some motors?

3. What is a synchronous motor?
4. What part of the motor is the stator and what is its purpose?
5. What is the purpose of the shaded pole in a small induction motor?
6. What is the purpose of the centrifugal switch in a capacitor-start motor?
7. Why does an induction motor have a certain amount of slip?
8. Explain how you would reverse the direction of rotation on a motor.
9. Explain the method to test for a winding-to-stator short-circuit on a motor.
10. Explain the method of testing the winding on a motor.
11. What is a continuity test?
12. How is a clamp meter used to measure line voltage?

Chapter 14

System Troubleshooting

INTRODUCTION

Troubleshooting and repairing a system requires knowledge and time. The greater the knowledge of a system and its associated components, the less the time required to find a trouble and make a repair. Troubleshooting often becomes a logical process of elimination.

The majority of the problems can be attributed to failure of electrical components within the control circuits. A technician must understand the operation of these electrical devices, and be able to follow their function and their sequence of operation from wiring diagrams.

When a technician is called to repair a system, he or she must be aware that time is of the essence. For example, the down time on a system, such as a supermarket freezer, could cost the customer thousands of dollars in losses. The timely restoration of a system and a satisfied customer is the best advertisement a service company can have.

Many troubles in complicated systems can be addressed with simple solutions. For this reason, a thorough evaluation should be made of the symptoms before trying to located the problem. It is good practice to make a list of observations and possible solutions. Veteran environmental control technicians have a saying about troubleshooting, *KISS*—which means *keep it simple, Stupid.*

By following these suggestions, problems involving a fuse, circuit breaker, compressor motor, or thermostat should be discovered immediately. Usually however, a blown fuse or tripped circuit breaker are symptoms of a problem rather than solutions. For this reason, after a repair the entire system must be tested for proper operation. Otherwise, expect a quick call back.

CIRCUIT DIAGRAMS

Refrigeration, air conditioning and heating manufactures usually supply electrical diagrams with each system. The diagrams are generally supplied in the installation and operation manual. Often electrical diagrams are placed on the rear of the main unit or on a removable panel.

Manufacturers may supply block diagrams, wiring diagrams, or schematics of the system. The operation manual often offers operation sequences and troubleshooting charts. Manufacturers will usually supply replacements if these instructions are lost. However, time is usually of the essence in the repair of a down system, and a technician often must develop a pictorial or block diagram of a system.

When developing a block diagram, every major component and sub system is drawn in their relative positions. Every known component is identified and all terminals are labeled. Main power and low voltage lines are traced and their colors noted. Often the problem can be pinpointed and repaired from a block diagram. If this is not possible, a more complete wiring diagram can be developed in the form of a ladder diagram.

LADDER DIAGRAMS

The sequence of operation of a system is difficult to follow with a typical electrical schematic. For this reason most manufacturers present the wiring diagram in the form of a *ladder*. Ladder diagrams identify the order of operation of devices, independent of their physical location.

The vertical lines in a ladder diagram represent the line voltage. These lines can represent AC or DC voltage of any amount. In a single-phase system the lines are labeled L_1 and L_2. In a three-phase system the vertical lines are labeled L_1, L_2, and L_3. The horizontal circuits represent the series and parallel circuit connections of switches, relays, and loads of a specific circuit.

The simple circuit in Figure 14-1 is a basic ladder diagram. The motor can be turned on by either of the parallel switches, S1 or S2. However, both switches must be off to disconnect the motor. The switches and the motor form a series-parallel circuit.

In the circuit in Figure 14-2 the switches are connected in series, and both must be closed to energize the motor. On the other hand, opening either switch will disconnect the motor from the line voltage. The switches

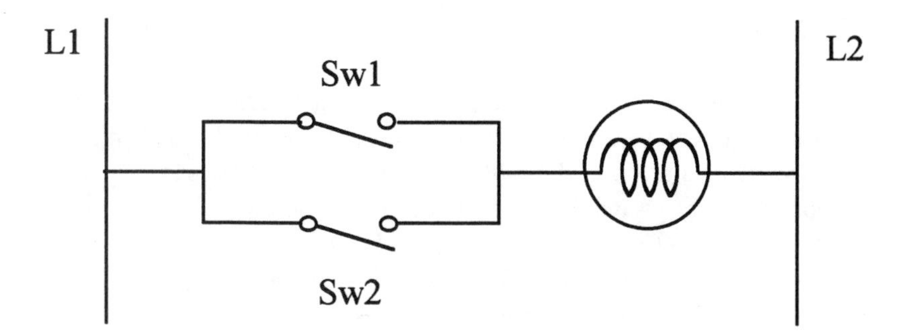

Figure 14-1. A ladder diagram of a series-parallel circuit.

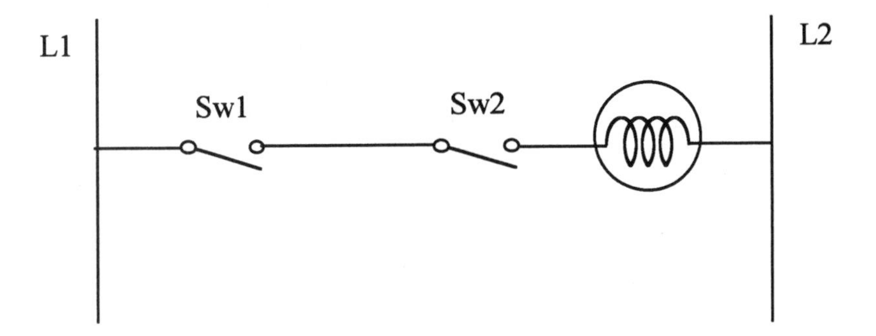

Figure 14-2. A ladder diagram of a simple series circuit.

and the motor form a series circuit. In most circuits the switches in this example would be replaced with relay or contactor contacts.

Correctly drawn ladder diagrams follow specific logic:

- The diagrams are read from left to right and top to bottom.
- Relay and contactor contacts are shown in the deenergized position.
- Contactor and relay contacts have the same numbers or letters as the coil.
- A contactor or relay may have both N.O. and N.C. contacts. However, each set of contacts changes condition when the relay is energized.

FREEZER WIRING DIAGRAM

Figure 14-3 is a ladder wiring diagram of a typical upright freezer. The circuit between points 1-1 is that of the cabinet door light. The push-button switch is shown in the OFF position with the door closed. When the door opens the switch closes and completes the light circuit.

The circuit between points 2-2 is an LED used as a signal light to indicate that the freezer is on. The LED is connected directly across the 120-V line in series with a current-limiting resistor, and is on whenever power is connected to the freezer.

The circuit between points 3-3 is a stile heater connected directly across the 120-V line. This resistance is in the circuit at all times.

The compressor motor and its control components are connected between points 4-4. From left to right, the circuit is comprised of a series thermostat, a series heat-sensitive overload device, and the parallel motor starting and running windings. The starting relay coil (R) is in series with the running winding. The N.O. contacts of the starting relay are in series with the starting capacitor and the top running winding and the starting solenoid. The series thermostat switch and the overload protection device must be closed for power to be applied to the compressor motor.

When the freezer temperature reaches a preset level, the thermostat makes a demand for cooling. The contacts close, applying power to the compressor. A sudden large current flows through the running winding and the series coil of the starting relay. This energizes the starting relay, closing the contacts in the starting winding. These contacts close and allow current through the starting winding, and the starting capacitor. When the motor reaches running speed, current decreases through the run winding and the relay coil, allowing the relay starting contacts to open. This removes the starting windings from the circuit.

TROUBLESHOOTING A FREEZER

The upright freezer in Figure 14-3 is a good basic refrigeration system for an introduction to troubleshooting. Evaluation of the system should be by observation, voltage and ohmmeter tests. When making observations all the senses such as sight, smell, hearing, and touch must be utilized.

Figure 14-3. A wiring diagram of a simple freezer.

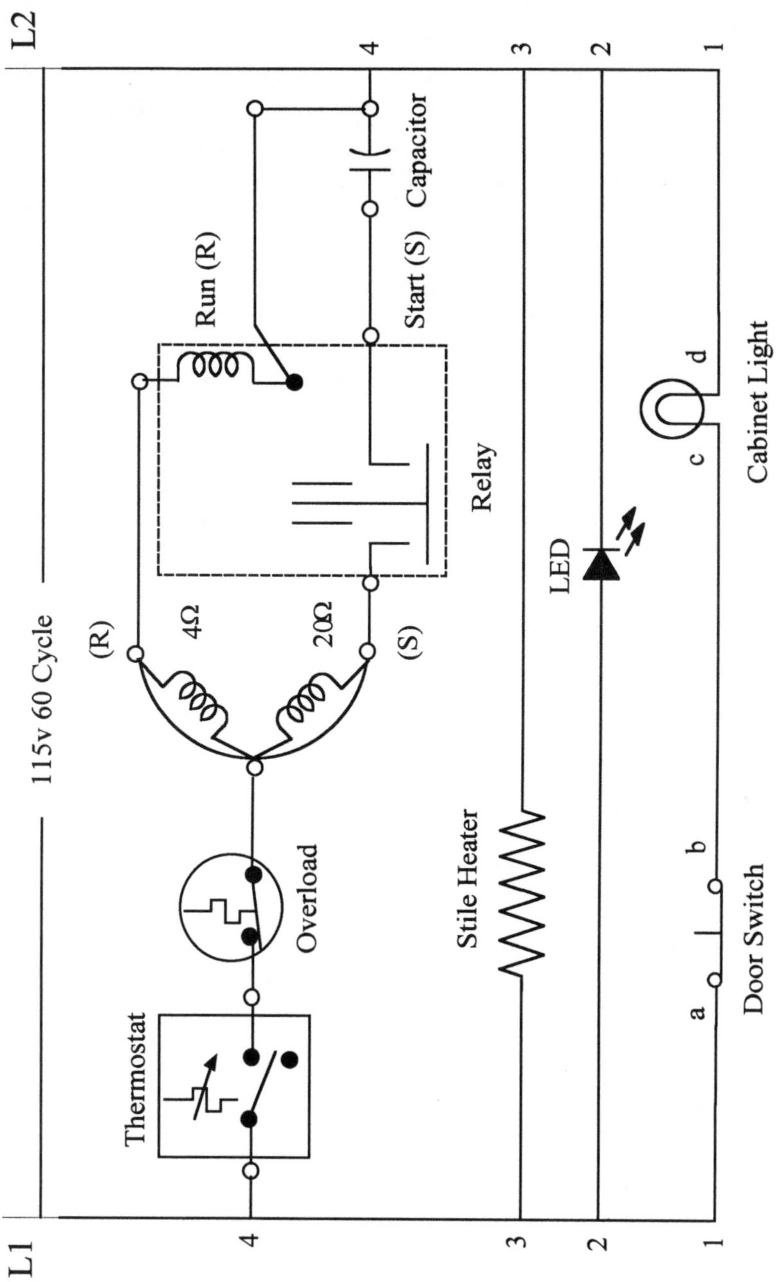

The conditions within the freezer should point to the problem by answering questions of existing conditions. For example, is the freezer temperature too low or too high, does the compressor run, or is it short cycling? The answer to these questions should localize the problem to a specific area of the circuit.

Most tests to localize faulty components may be performed with a voltmeter or clamp-on ammeter. Tests on faulty components can often be performed with an ohmmeter. Power must be disconnected for any test with an ohmmeter. Suspected components that cannot be tested in the field must be substituted. For example, a faulty capacitor cannot always be detected by an ohmmeter.

Table 14-1 is a general troubleshooting and repair table for the freezer in Figure 14-3. The symptoms of problems are taken upward from the bottom of the diagram.

The fault in Complaint #1, *no cabinet light*, can be detected in several manners. Always eliminate the most probable cause or make the easiest, quickest test first.

1. Assure that there is power to the freezer.

2. Replace the light or test the resistance with an ohmmeter.

3. Make an ohmmeter measurement across the door switch with power off. A reading near zero is correct.

4. Disconnect power, jumper across the door switch and turn the power on.

5. Look for loose wires or connections.

6. An ohmmeter can be used in a bridge test from L_1 to point "a" through point "d" and to line 1. The reading from "a" through "c" should read near zero. Readings to points "d" and L_2 should read the low resistance of the bulb.

7. Finally, turn on the power and bridge a voltmeter from line 1 to points "a" through "d" from left to right. *Each reading will require the power to be restored*. An open connection will cause a 120-V reading.

COMPLAINT	POSSIBLE CAUSE	SOLUTIONS
1. No cabinet light	Faulty bulb Open switch Loose wire	Replace Replace Correct
2. Compressor doesn't start—no hum.	No power Thermostat Broken or open wire Open starting control Open motor winding	Restore power Check thermostat Repair Replace Replace motor
3. Compressor doesn't start—trips overload protection device.	Low voltage Defective starting relay Defective starting capacitor	Restore voltage replace Test and replace
4. Compressor short cycles.	Low refrigerant Low voltage High discharge pressure	Recharge Restore voltage Ventilation/clean condenser, check charge
5. Compressor runs—draws high current and overheats	Low voltage Low voltage Shorted starting relay	Restore voltage Restore voltage Replace
6. Compressor short cycles	low refrigerant pressure Excessive current	Recharge Check motor current

Table 14-1. Basic troubleshooting of an upright freezer.

The fault in complaint # 2, *compressor doesn't start—no hum,* must be solved by a number of methods. We can begin by eliminating the start capacitor, and either the start winding or the run winding. If only one of these three were defective there would be motor hum.

Start with the easiest possibility.

- Use a voltmeter to determine if there is power to the motor circuit and motor terminals.

- Check that the thermostat is closed. If necessary jumper the thermostat contacts.

- Look for broken or loose wires.

The freezer circuit may be tested *live or dead*. Live tests may be performed with a voltmeter, but never with an ohmmeter. The voltmeter can be bridged or hopscotched across components in a series circuit. The ohmmeter can be used on a dead circuit to test for opens, shorts, continuity or changes of resistance.

When performing a hopscotch test along a series circuit, a reading of line voltage indicates an open connection or component of the last device bridged. The following is a summary of these tests on the freezer in Figure 14-3.

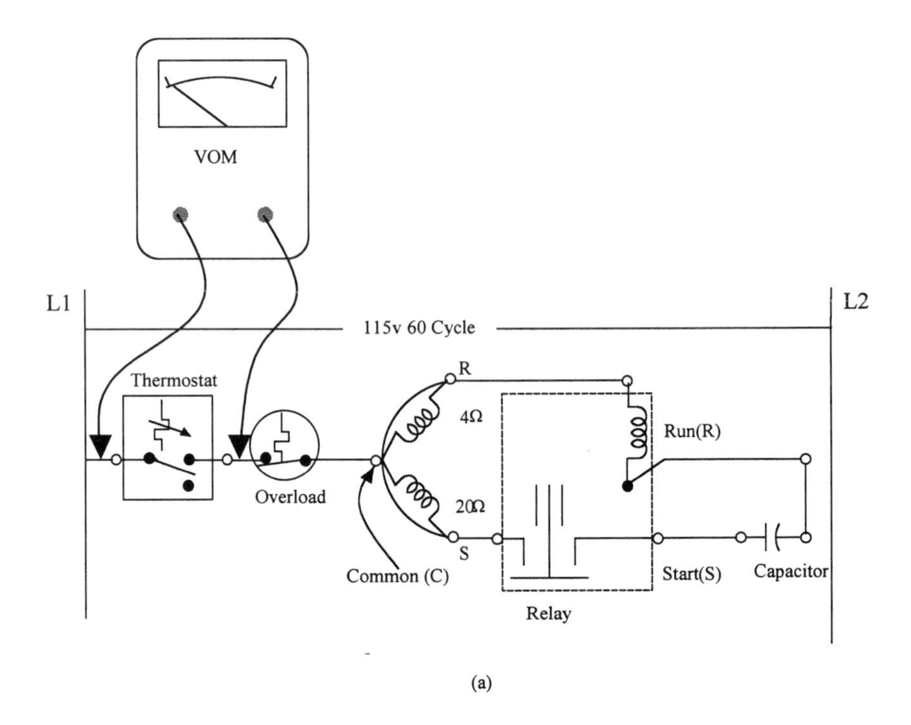

Figure 14-4. Bridge testing the motor circuit with a voltmeter: (a) from line across thermostat; (b) from line across overload device.

OHMMETER TEST

1. A bridge or hopscotch test is a step-by-step test performed with a voltmeter, or with an ohmmeter. A reference point is established and the instrument is placed across each component. An ohmmeter or a voltmeter reading is made from the reference point. The readings are evaluated to determine if the component is shorted or opened. Power must be off to use the ohmmeter. To perform the test, bridge the meter across each component or part of the circuit. Opening and closing the thermostat should read zero and infinity. Across the overload device should read zero. Across the motor terminals C to R should read the resistance of the run winding, approximately 4 ohms. A high reading indicates an open run winding. Across the motor terminals C to S should read approximately 20 ohms. A high read-

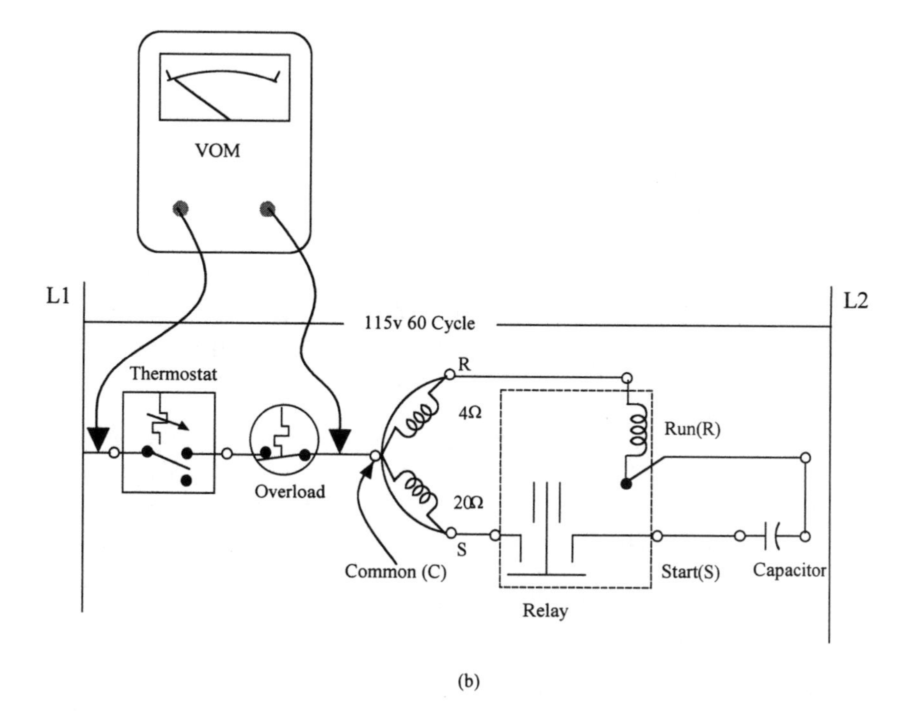

Figure 14-4. (Continued).

ing indicates an open starting relay or an open starting winding. Any excessively low or high reading is cause to replace the compressor motor. Use proper recovery methods.

2. With power removed, an ohmmeter can be used to repeat the bridge test. The test can be performed step-by-step as with the voltmeter, or across each component individually. Opening and closing the thermostat should read zero and infinity. Across the overload device should read zero. Across the motor terminals C to R should read the resistance of the run winding, approximately 4 ohms. A high reading indicates an open run winding. Across the motor terminals C to S should read approximately 20 ohms. A high reading indicates an open starting relay or an open starting winding. Any excessively low or high reading is cause to replace the compressor motor.

   When the compressor is replaced, the proper reclamation methods must be employed and the compressor recharged and tested.

VOLTMETER TEST

Voltage readings indicated here are for an assumed line voltage of 120-V AC. Normal line voltage can be several volts less than 120-V.

1. With power on, connect common meter lead to L_1. Connect the red lead to the right side of the thermostat. The meter should read 120-V with the thermostat open and zero with the thermostat closed. A reading of 120-V for both tests indicates an open thermostat.

2. Assure that the thermostat is closed, and move the red lead of the meter to the right side of the overload device. A reading of 0-V indicates the thermostat and the overload device are functioning correctly, or that there is an open between the red meter connection and line L_2. A reading of 120-V indicates that the overload device is open must be replaced. The motor must be replaced if the overload device is internal.

3. Move the red lead to point R on the right side of the motor. A reading of zero indicates an open wire at the connection to L_2. A reading of 120-V indicates only that there is power between L_1 and L_2.

4. Move the common meter lead to Point C on the motor and the red

lead to point "S." A reading of zero indicates that the capacitor is open. A very low reading indicates that the capacitor has lost capacitance. A reading of 120-V line voltage indicates that either the starting winding or the relay contacts are open and the motor must be replaced.

TROUBLESHOOTING A GAS FURNACE

Natural gas or propane are the most common heating fuels. Over 250,000 gas heating systems are sold in the US every year. Gas heaters and furnaces have a number of basic safety devices. The control valves necessary to safely operate a gas heating system are:

1. A *manual shut-off valve* is required on every gas appliance. The function of this valve is to remove gas from the main burner. A secondary valve is often used to turn off the gas to the pilot.

2. A *pressure regulator* is used to control the gas pressure and to deliver the correct pressure to the burner manifold. The function of the pressure regulator is to maintain a relative constant output pressure in spite of changing input pressure. Most pressure regulators can be converted from natural gas to LP gas with a simple valve change. The pressure should be then adjusted for fine tuning.

3. The *ignition system* is to light the gas at the heat manifold. There are three methods used to produce ignition.
   - The *standing pilot system* has a gas pilot that remains on at all times.
   - The *intermittent pilot system* lights the pilot upon heat demand. The main burner is subsequently ignited.

4. The *direct burner ignition system* lights the main burner without the use of a pilot. This is accomplished by one of two methods.
   - by a spark from a spark generator,
   - by a hot surface igniter that is heated by electricity.

To eliminate the need for a continuous pilot and to conserve energy, most new furnaces will be operated by a spark generator or a hot surface igniter. The spark generator and the hot surface igniter are operated by a microprocessor in newer systems.

A *safety lockout valve* shuts off the gas to the main burner and the pilot if the burner flame fails.

The *automatic On-Off valve* is controlled by the thermostat on a demand for heat. In some systems the safety lockout valve and the automatic On-Off valves are combined into one unit. The option is up to the manufacturer.

The *over-temperature control* shuts off the gas to the pilot and main burner in the event that the system overheats. This control is usually located in the plenum or the warm air ducts. Some over-temperature controls are snap type switches that must be manually reset.

New heating systems employ a *combination gas control valve* (CGC). The valve normally includes:

1. A fail-safe lighting gas cock that prevents the main burner from igniting as the pilot is being ignited.

2. An adjustable servo pressure valve for better regulation. The pressure may be set for a manometer to compensate for pressure requirements, such as might occur at different altitudes.

3. Pilot gas filters and flow adjustment.

An example of a gas control valve that integrates these functions is shown in Figure 14-5.

Figure 14-5. A gas control valve. (Courtesy, White-Rogers Div. Emerson Electric)

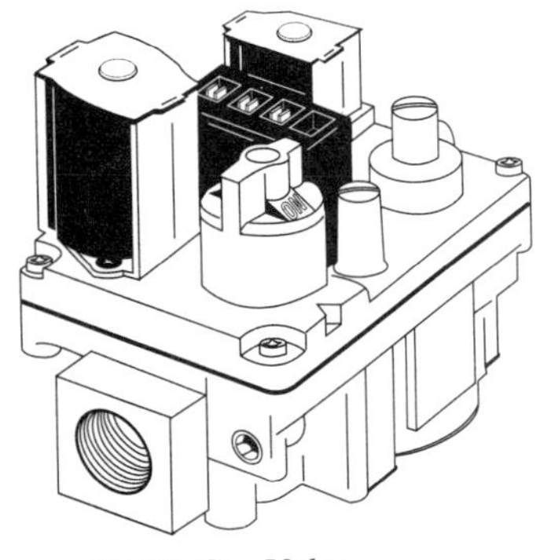

36E01 Gas Valve

Thermocouples play a major role in the control of safety valves in heating systems. As was noted in Chapter 11, the thermocouple is comprised of two dissimilar metals which are formed into a junction. When the junction is heated a voltage potential difference is developed, much as if the junction were a battery. The thermocouple must have sufficient output voltage and current to activate the valve that it is controlling.

The safety shut-off valve is controlled by a relay called a Pilotstat™ (Figure 14-5). The voltage generated by the thermostat or thermopile is applied to the electromagnet. However, there is insufficient current to fully activate the valve plunger. The reset button on the combination gas control must be pressed, to reset the control. A normal flame and thermocouple produces enough energy to keep the plunger in the holding coil. Figure 14-6 illustrates a basic thermocouple safety shut-off circuit.

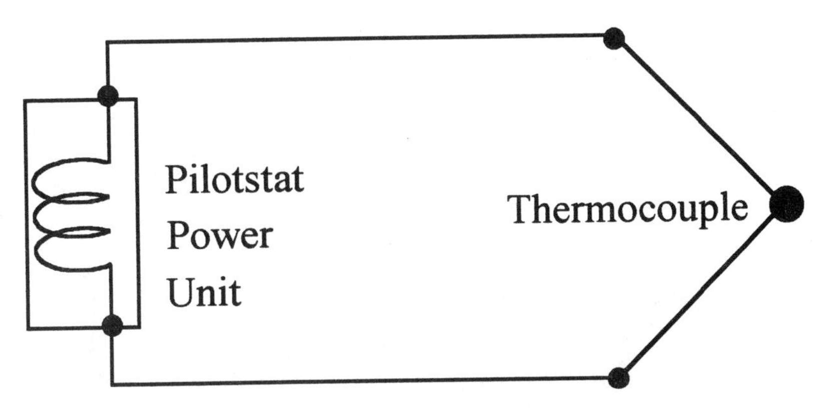

Figure 14-6. An example of a thermocouple safety circuit.

The pilot and the thermocouple are the primary safety devices in many types of heating systems. There are numerous factors that can cause pilots faults. Some of these are given in Table 14-2. The color, shape, and size of the pilot give a clear indication of the causes of system shutdown.

1. The thermocouple or thermopile is the energy source which keeps the main gas control valve energized and open. Decrease or loss of voltage from the thermocouple unit results in deactivation of the main burner valve. Failure of a thermocouple must be ascertained by proper troubleshooting methods. Once there is a strong suspi-

PROBLEM	POSSIBLE CAUSE	CORRECTION
Pilot will not light	Main gas valve turned off	Turn on valve
	Pilot supply turned off	Turn on valve
	Air in main gas line	Purge line
	Pilot reset button not engaged	Engage
	Control valve not in pilot position	Correct
	Pilot burner clogged	Clean orifice
Pilot will not hold when reset button released.	Reset released too soon	Restart
	Improper pilot flame	Adjust flame
	Faulty power unit	Replace
	Pilot burner clogged	Clean orifice
	Loose electrical connection	Correct
Pilot on, main burner off	Gas line restricted	Correct
	Powerpile fault	Replace
	Incorrect pilot flame	Adjust
Improper gas pressure	Test Pressure	
	Faulty thermocouple	Replace
Loss of pilot	Pilot unshielded from draft	Shield
	Clogged pilot filter	Clean, replace
	Improper pilot flame	Adjust
	Faulty powerstat unit	Replace
	Low gas pressure	Adjust pressure
Faulty pilot flame	Pilot pressure high	Adjust
	Pilot burner orifice too small	Replace
	Pilot burner orifice too large	Replace
	Pilot air opening clogged	Clean
	Pilot filter clogged	Clean, replace
	Gas pressure low	Adjust

Table 14-2. Pilot Fault Analysis

cion of thermocouple, the unit is usually replaced, rather that making further tests, due to the cost of technician labor. The first step in the evaluation of a thermocouple is to determine that the thermostat is activated. Possible problems with the thermostat can be eliminated by jumpering the thermostat terminals at the lead terminals in the furnace. If the system operates correctly with a jumper in place the thermostat is misadjusted or faulty.

The standing pilots are being replaced in new furnaces with other methods of ignition. An example is the hot surface ignition system shown in Figure 14-7. The furnace is a low-exhaust-temperature system, controlled by the thermostat and a solid-state controller. Before gas is applied to the hot surface igniter the following events take place:

1. The thermostat demands heat.
2. The controller checks for open limit contacts. If a limit contact is open the appliance will remain inoperable until the problem is corrected. The status light will blink a code to indicated the particular open contacts.
3. The controller checks to assure that the vent pressure switch is open.
4. The vent blower is energized.
5. The vent pressure switch closes and closure is detected by the controller.
6. Flame rollout switches are checked to assure they are closed.
7. The blower purges the vent for 15 seconds and the hot surface is ignited.
8. After 17 seconds hot surface warm up, the main valve opens.
9. The flame burner ignites and the flame sensor is activated. The igniter will de-energize in about 8 seconds. If after 4 attempts the sequence is not completed the controller will turn the system off for approximately 1 hour, unless the room thermostat is reset.
10. Thirty seconds after the main valve is energized the heating speed of the air blower is activated.
11. The furnace remains in operation until the room thermostat is satisfied.
12. The air blower will remain on for approximately 3 minutes to cool the heat plenum.

TROUBLESHOOTING AN AIR CONDITIONING SYSTEM

Troubleshooting the electrical circuits of a 240-V or 440-V air conditioning system can usually be separated into two areas; the high voltage circuits and the low voltage control circuits. The high voltage supplies the low voltage, and without the high voltage there will be no low voltage. Therefore, the first steps are to assure that line voltage is present to the unit, and within the specifications' limits.

Figure 14-7 is a wiring diagram of an air conditioning and heating unit. Before beginning troubleshooting we must become familiar with the wiring and the controls of the entire system.

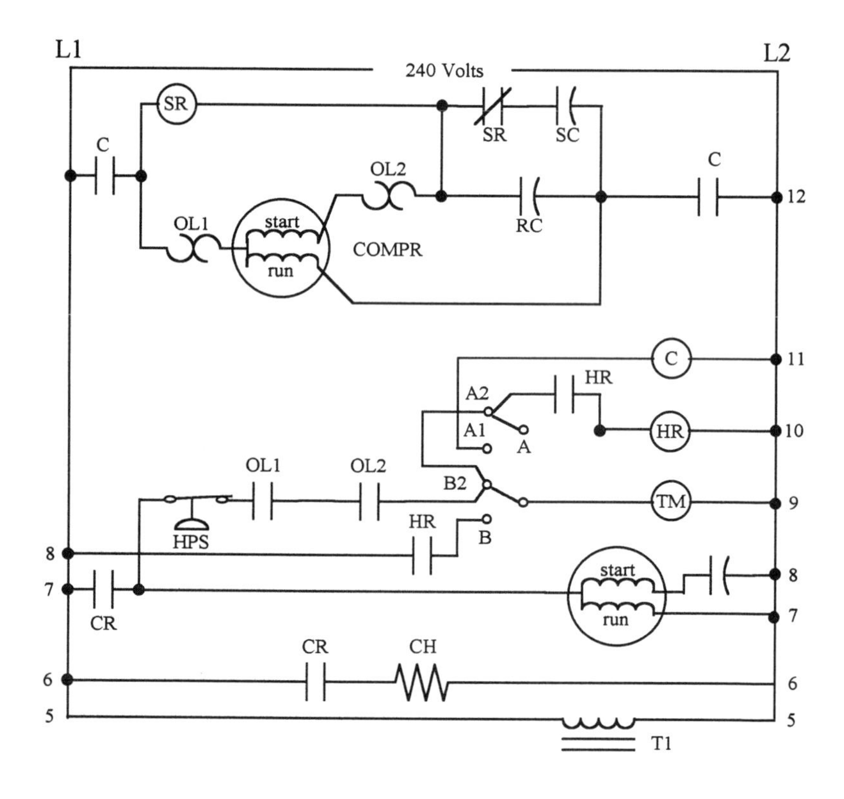

Figure 14-7. A wiring diagram of a heating and air conditioning system.

BASIC CONSIDERATIONS

1. All controls shown on the wiring diagram are in the off position.

2. The diagram is divided into three parts: the three-phase air conditioning unit, the common thermostat, and the heating unit.

3. The diagram is also divided into the low-voltage control circuits and the high-voltage circuits.

4. The system compressor uses a single-phase 230-V capacitor-start capacitor-run motor.

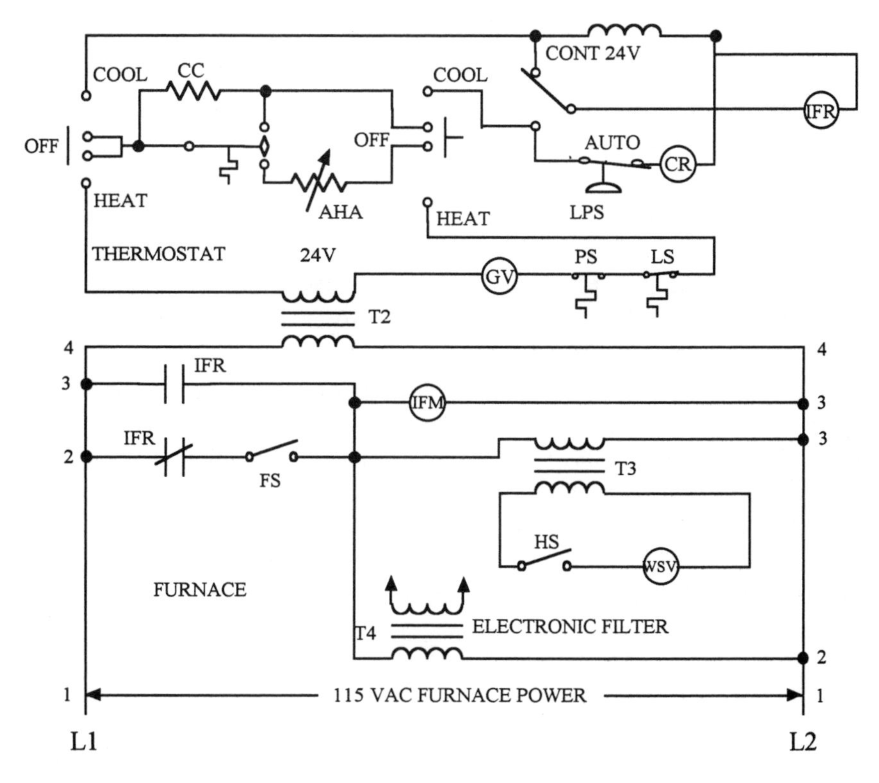

Figure 14-7. (Continued).

5. The system fan is common to both the heating and cooling system.

6. The system fan is powered by a single-phase, 115-V AC motor controlled from an indoor fan relay shown in the thermostat.

7. All low-voltage circuit components are shown at the bottom of the wiring diagram.

8. The common thermostat is shown in the center of the diagram.

9. All high-voltage circuit components are shown at the top of the wiring diagram.

10. The system humidifier and an electronic filter are shown at the bottom of the diagram and are powered from the 155-V AC furnace supply.

Many of the system's components are interrelated; nonetheless, it must be analyzed as several separate units.

The Thermostat

The thermostat controls both the heating and air conditioning systems, and continuous fan control for air circulation. In the HEAT position the thermostat switch completes the gas valve (GV) circuit through the pilot safety relay (PS) and the heat limit switch (LS). The thermostat contains energy-saving components in the form of an adjustable heat anticipator (AHA) and a cooling anticipator (CC).

The thermostat should be the first suspect, after power, for any system where the heating, the air conditioning or both are completely down. The thermostat can be isolated by jumpering COOL-HEAT terminals. Both sets of terminals must be shorted. For example, to test the furnace the HEAT to COOL terminals must be jumpered at both locations on the thermostat switch. However, thermostats seldom fail and are usually replaced rather than repaired.

The Humidifier

Activation of the humidifier is controlled by the humidistat (HS) and the water solenoid valve (WSV). Power to the humidifier is supplied by closing either the fan switch (FS) or the Indoor fan motor relay (IFM) to apply the 115-V AC through transformer T4.

The Electronic Filter

The electronic filter is powered by closure of either the fan switch (FS) line 2 or the N.O. indoor fan motor contacts (IFR). The IFR contacts are controlled by the IFR relay in the thermostat.

Table 14-3 is a summary of the symptoms of problems in the heating unit, the electronic filter and the humidifier.

PROBLEMS	POSSIBLE CAUSE	CORRECTION
No heat	T4, open PS, or LS Also see Table 14-3	
Faulty Electronic Filter	Thermostat off Loose connections or faulty T4.	Correct, replace T4
Faulty humidifier	Fan switch (FS), or indoor fan relay.	Check switch, relay And connections. Repair or replace

Table 14-3. Heating, humidifier, and electronic filter problems.

The Heating System

The heating system is supplied by a 115-V AC line. The heat-cool switch supplies a heat demand. Most probably the system would be controlled by a automatic heat-cooling thermostat or a microprocessor. Line voltage L_1 and L_2 supply voltage for the electronic filter transformer T4, the humidifier transformer T3 and the gas valve transformer T2. The humidifier and the electronic filter are common to heating and air conditioning systems.

Air Conditioning Unit

The air conditioner in Figure 14-7 operates on 240-V AC and is controlled by the COOL switch on the thermostat. The compressor uses a capacitor-start capacitor-run 230-V motor. The indoor fan relay (IFR) is supplied power through T1 from the 230-V AC source in the continuous operating position.

Examine the ladder diagram of the air conditioning system from bottom upward. *Each circuit is connected between supply voltage lines L_1 and L_2.*

Points 5-5—Transformer (T1) supplies the 24 volts for the thermostat circuitry in the COOL mode of operation. Power is applied to transformer T1 whenever the 230-V line voltage is present.

The run windings of the compressor are energized from lines L_1 and L_2 through contacts C on each side. Current flows through the run capacitor (RC), the run winding and the overload relay contacts (OL_1). Note that, for safety, power to the compressor is removed from both lines by the N.O. contactor relay contacts.

Points 6-6—The crankcase heater (CH) is in series with the N.C. contacts of the control relay (CR). Contacts CR open when the compressor contactor is energized, disengaging the heater.

Points 7-7 & 8—is a series-parallel circuit with the N.O. contacts of the control relay (CR) in series with the parallel start and run winding of the fan motor (FM). The fan motor is a capacitor-start capacitor-run single-phase induction motor.

Points 7-9—is a series circuit consisting of the N.O. contacts of the control relay, the safety contacts of the high pressurestat and the two overload device contacts (OL_1 and OL_2), the N.C. relay contacts (A1 and A2), and the timing motor. The timing motor begins a 15-second delay when control relay contacts (CR) close. At the end of the 15 seconds, relay contacts A and A1 close and energize the compressor contactor coil (C) through contacts (HR), contact B2, OL_2, OL_1, HPS, and CR.

Energizing the compressor contactor (C), points 9-1, connects 230 volts across the compressor motor. This energizes both the starting circuit and the run circuit. Current in the run winding flows through the run winding and the contacts of the overload device (OL_1).

Current in the starting winding flows through the overload device contacts (OL_1), the start winding and overload device contacts (OL_2), and splits to flow through the parallel circuit comprised of the starting relay contacts and the starting capacitor (SC) in parallel with the run capacitor (RC). When the motor nears running speed the contacts of the starting relay open to remove the starting capacitor from the circuit. The starting winding remains in the circuit but at a lower current through the run capacitor (RC).

Faults with an air conditioning unit can be localized by the following troubleshooting techniques. Use your basic senses. Table 14-4 is a summary of some typical problems and their solutions.

1. First make sure that main power is applied to the system.

2. Smell for overheating or burned wiring insulation.

3. Look for loose connections, burned insulation, loose parts, or dampness.

4. Listen for unusual noises, and overcycling.

5. Touch the compressor and transformer cases to feel for excessive heat.

Noise in an air conditioning system may usually be localized by hearing or feel. A useful and inexpensive device is an automotive stethoscope that can be used to pinpoint noise.

When testing and repairing a system and all else has failed, it is appropriate to use a substitution test. When substituting a part in a system there are several rules that must be followed.

- Substitution is performed as the last resort.
- The part substituted must be the same as the part being replaced.
- All safety precautions must be observed.

TROUBLE SHOOTING A THREE-PHASE AIR CONDITIONER

The techniques used to troubleshoot three-phase systems are the same as for single-phase systems. The exceptions are problems that may

PROBLEM	POSSIBLE CAUSE	SOLUTION
Compressor inoperable	No power	Return power
	Blown fuse	Replace
	Open circuit breaker	Reset
	Open wire	Repair
	Open contactor	Replace coil, or contactor
	Transformer bad	Replace
	Thermostat open	Correct
	Bad thermostat	Replace
	Thermal overload	Test, replace if possible
	Internal thermal overload defective	Replace compressor
	Pressure switch open	Test, replace
Compressor short cycles	Condenser air flow	Correct
	Condenser fan	Repair or replace
	Refrigerant undercharge	Correct
	Refrigerant overcharge	Correct
	Air in system	Evacuate and recharge
	High-pressure switch open	Check charge, clean condenser
	Starting capacitor	Replace
	Running capacitor	Replace
	Low line voltage	Correct
	Time-delay circuit	Repair or replace
	Dirty air filter	Clean or replace
	Frosted evaporator	Check suction pressure
Insufficient cooling	Low suction pressure	Correct
	High suction pressure	Correct
	Low-pressure switch defective	Replace
	Low-pressure switch misadjusted	Adjust correctly
	Evaporator fan	Test, replace
	Evaporator fan relay	Test, replace
	Air filter dirty	Clean or replace
Excessive cool	Thermostat	Adjust, replace
	Shorted contactor	Replace
	Shorted pressure control	Replace

Table 14-4. Air conditioner problems.

occur with three-phase motors, and are discussed in Chapter 12 under that heading

SUMMARY

Troubleshooting electrical systems in heating and air conditioning systems requires the understanding of the electrical controls and their function in the system.

Troubleshooting is a mystery-solving process. A set of known facts is used to discover clues and to finally make a logical decision.

QUESTIONS

What is a bridge or hopscotch test?
When is an ohmmeter used to troubleshoot a system?
When is a substitution test performed on a system?

Safety

A primary concern of both the technician and management must be safety—personal safety and safety to the system. Of course, personal safety always takes precedent over installation, maintenance, and/or repair time.

The human body functions by way of small electrical signals from the brain. The heartbeat is triggered by these same small electrical signals. This means that the body will react adversely to any outside electrical signal. We call this effect on the body is a shock. Figure A-1 is a table of the body's responses to electrical current.

The amount of current that will flow through the body is controlled by Ohm's law. This means that the current flow depends upon the magnitude of the voltage contacted and the resistance of the body between the contact points. The resistance of the body depends upon the amount of moisture on the body, the resistance of the contact area. For example, A short from the hand to ground would have very high resistance if the person were wearing rubber-soled shoes and standing on a dry surface. The resulting high resistance would produce little current flow and only a slight shock. On the other hand, a person leaning his or her left hand on a grounded surface and touching the right hand to 240 volts

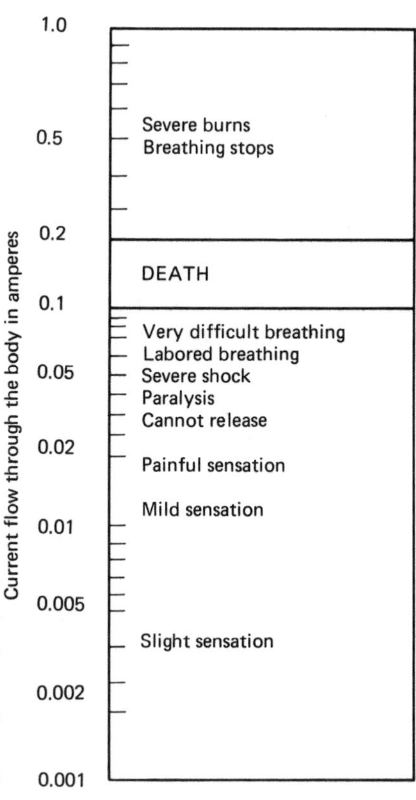

Figure A-1. Table of current effects on the human body.

would probably be electrocuted. This is because the current would pass through the chest area from one arm to the other. The same would apply to someone standing on a wet surface and toughing 240 volts.

The bottom line is that, under the right conditions, a shock from a very small voltage may be fatal. Therefore, never take even the slightest shock on purpose.

The following pages are a summary of safety rules. These rules should be reviewed even by experienced technicians. There is an old saying about electricity, "familiarity breeds contempt." This means, as we become more experienced, we tend to become more relaxed about safety. After all, we have worked on active circuits many times without being shocked, or we have been shocked many times without a serious injury. Just remember it takes only one minute of inattention to be seriously injured or killed.

Safety for yourself, your co-workers, and equipment are the most important conduct that you can acquire in your goal to become a skillful technician. The purpose of this summary is to provide you with essential instruction in personal safety, and the safe usage of tools and equipment. Emphasis is given to the concept of creating safety awareness as part of your daily life. We believe that our effective ability to install, repair, and maintain air conditioning, refrigeration, and heating equipment is largely dependent on the extent that safety thoughts become part of our daily behavior. For this reason we present you a number of safety thoughts. Study them carefully, then take the safety test. When you have answered all the questions correctly, you can be confident that your behavior is beginning to represent and element of safety awareness.

ELECTRICAL SAFETY

The most important thing to remember about electrical shock is that the current rather than the voltage determines the shock intensity. Of course, voltage must be present to cause current, but the resistance of the human body and the contact resistance can vary from a few ohms to many thousands of ohms. Since current kills, it is important to make sure that your body always has a high resistance between an electrical circuit and ground.

When the skin is dry, it presents a high resistance of several thousand ohms to a voltage. However, when the skin is damp or wet the contact resistance can be less than 1,000 ohms, in which case a low voltage can cause a fatal current to exist. The actual body contact resistance depends upon the area of contact and the wetness of that area. For that reason it is important to never lean on a piece of equipment or a ground, and always wear rubber-soled shoes.

An electrical shock to a technician is usually the result of the shorting of some part of the body between a voltage and ground. Ground may be any metal part of the equipment, a gas pipe, water pipe, or the earth. Voltage-to-ground shock is the most common; however, shock may occur between two pieces of equipment. This type of shock is the most common.

The best single rule to follow is never take a chance on any value of voltage in any circuit. The following safety precautions should always be taken when working with electrical equipment.

- Never touch even a "dead" circuit or line with an open hand, as the muscles will cause the hand to clamp on the circuit. Use the back of the fingers for the final "hot" test before working on a circuit; even though the voltmeter reads zero volts.

- Never touch even one wire of a live circuit.

- Never work on electrical equipment alone.

- Turn off all power, and ground all voltage points. Make sure that the power cannot be accidentally restored.

- A capacitor can store a lethal charge.

- Never work on live equipment when tired, and always keep one hand in a pocket when working on live equipment, so that any shock will not pass through your chest and heart.

- Move slowly, make sure that your feet are placed for good balance. Never lunge for falling tools.

- Do not handle electrical equipment that is not grounded, unless it is triple insulated UL approved.

- Put out electrical fires with CO_2 or an approved extinguisher. Do not use water.

- Never touch two pieces of electrical equipment at the same time. One might have an electrical voltage present.

- Wear rubber-soled shoes to prevent grounding of your body in the event of an electric shock.

- Work on electrical equipment only when you are physically fit and provided with the proper equipment, tools, and safety devices. Contact your supervisor when anything is wrong.

- Avoid interfering with or startling anyone working on a live electrical circuit.

In summary the number-one rule on working on electrical equipment is NEVER WORK ON A LIVE PIECE OF EQUIPMENT. Having said that, we know that it is an impossibility. There are times when the only way to find a trouble is with the equipment on and fully operable. For those times, always take extreme caution and apply the preceding rules with care.

GENERAL SAFETY

There are many factors other than electricity that must be considered to ensure safety to the student or work and tools and equipment. Study the following rules carefully to ensure safe working conditions for you and your colleagues.

- Wear snugly fitting clothing. Never wear loose clothes, long sleeves, dangling neckties, loose trouser cuffs, finger rings, or other unsafe apparel when working around machinery.

- Wear goggles when doing any work where splashing or spraying material may enter the eyes.

- Do not exchange your mask or goggles with another person without having them sterilized.

- Always wear rubber-soled shoes when working on electrical equipment on the floor.

- Take care in using an air hoses to clear dust from equipment. Wear goggles and never direct an air hose at another person.

- When lifting heavy objects use the leg and arm muscles and wear a back support. Keep your back as straight as possible.

- Be sure that your hands are as free as possible from dirt, grease, and oil when working with tools.

- Walk—do not run—in a work area.

- Use both hands when climbing up or down ladders. If this is impossible, climb one rung at a time and at a slow pace. Make sure that you are on the bottom rung before stepping off the ladder.

- Caution any other student or worker that you see violating a safety rule.

- Know where the fire extinguishing equipment is located and how to use it.

- Use only the proper type of extinguishing agent for an electrical fire. Put out electrical fires, grease fires, or plain fires with CO_2 or an approved foam extinguisher. NEVER USE WATER ON AN ELECTRICAL FIRE.

- Burns from chemicals (such as acid) should be washed immediately and thoroughly.

- Apply first aid to small cuts and scratches, as they may become infected.

- Use soaps—not oils, solvents, or other compounds—for cleaning your hands.

- Practice procedures to follow in case of fire, earthquake, or other disasters.

- Signs are time savers as well as accident savers. Read them, and follow their advice.
- Report any injuries promptly.
- When rendering first aid do not move the patient unless absolutely necessary until you are sure what the injury is and have given first aid as necessary.
- Keep fire doors, aisles, fire escapes, and stairways clear.
- Keep control panels and circuit boxes clear.
- Use the proper type and size of screwdriver for the work.
- Use the proper fastening tool to remove and replace equipment fasteners.
- Always use the correct tool. Don't force a tool to do a job for which it is not designed.
- Keep tools and materials from projecting over the edge of a work bench, and off the floor.
- Don't carry a screwdriver in your pocket.
- Pull on wrenches—don't push. Make sure your footing is secure and allow plenty of clearance for your fingers.
- Handles of hammers are for gripping only. Don't use a hammer as a pry bar.
- Don't use a file without a handle.
- Always apply first aid to an injury; the smallest injury can become infected and cause larger problems.
- Don't use pliers as a wrench.

- Always carry long objects, such as ladders, with the front end high enough to avoid striking anyone.

- Power should always be shut off after electrical equipment has been used.

- Be sure that your hands are dry before touching electrical switches, plugs or receptacles.

- Never bridge a fuse with a wire or increase the current rating of a fuse. This may cause a fire.

- When removing electrical tools from a receptacle, always pull the plug handle and never the cable. DON'T GIVE A JERK AND BE A JERK.

- All portable electrical tools and test equipment are to be disconnected when not in use.

- Report any odor about environmental equipment and clear the area. Never turn on light or electrical equipment when there is a gas odor present. Clear the area at once.

- Before plugging in any electrical tools or test equipment be sure that the power switch is in the off position.

- Make sure that all portable electrical equipment such as drills or test equipment is grounded.

- Remove the chuck key immediately after using it.

- Hold an electric drill firmly, but ease up on the pressure just before the drill breaks through the work piece.

- Be sure that material to be drilled is held firmly so that it cannot move if the drill bit binds.

- Never use power tools when the cord is in need of repair.

- Do not handle electrical equipment when either you or the equipment are damp or wet.

- Never clean equipment when the power is on.

- Never use makeshift or unsafe wiring as it may cause a shock or fire.

- Start artificial resuscitation at once, if required, after safely removing the victim from the live circuit.

SOLDERING

Soldering electrical wires and components requires the use of the proper type of heating device such as a *soldering iron* or *soldering gun*. Both are required to heat to over 500 degrees F to melt the solder and joint to be welded for an effective connection. Soldering irons come in several wattage and heating ratings. The proper iron should be selected for the job. Lower wattage irons are advised for use when soldering on printed circuit boards to prevent the copper from lifting from the board. Soldering guns are faster and have high heat ratings, and are effective for soldering large conductors and areas. However, the high temperature can damage printed-circuit boards, insulation, and semiconductor components. NEVER USE A SOLDERING GUN ON SEMICONDUCTOR DEVICES OR PRINTED-CIRCUIT BOARDS. The following rules along with common sense, should be employed when soldering.

- Lift a soldering iron, hot or cold, only by the handle.

- When soldering be sure that the iron is placed where it cannot be inadvertently touched.

- Keep a soldering iron in a holder when it is not in use.

- When desoldering wire, make sure that there is no tension on the wire that can cause hot solder to flip onto your face.

- Wipe excessive solder from the tip with a soft rag; never flip it.

- Place rags, Kim Wipes etc., in a location away from the soldering iron.

- When cutting wires, keep the open side of the cutter away from your body.
- Use a face shield or goggles when soldering.
- Never interfere with someone when they are soldering.

SAFETY TEST

To assure a safe environment for the student or technician and the test equipment, the following safety test should be taken after a careful study of the Safety chapter. A passing grade for the test should be 100 percent. Failure of the test requires that the technician review the chapter and retake the test.

QUESTIONS

Answer the following questions true or false or by selecting the proper word to make the statement true, or by circling the correct word or phrase.

1. When splashing material might enter the eye, you should always wear ________ (goggles, shoes.)
2. You may never borrow personal equipment before first having it _______. (checked out, sterilized.)
3. You should wear _______ when working around machinery. (safety glasses, loose clothing.)
4. The fellow in the figure (opposite) is properly dressed for working safely.
5. (T-F) Wear shoes sufficiently heavy to give adequate foot protection.
6. What should you do when a liquid is spilled on the floor? (leave to the janitor, wipe up immediately.)

7. When lifting always keep your back in a _____ position. (bent, straight)

8. Is it a safe practice to play with an air hose? (yes, no)

9. (True, False) Always keep your hands clean when working.

10. What should you do if you see another student or worker violate a safety rule. (ignore it, tell him/her)

11. (True, False) Practical jokes keep the shop a happy place and are a safe practice.

12. (True, False) Litter on the floor and work benches should be left for the janitor.

13. Can shop signs help prevent accidents? (Yes, No)

14. Cutting oil and solvent are safe products for cleaning your hands. (Yes, No).

15. _______ procedures to follow in case of a disaster. (Practice, Ignore)

16. Burns from chemicals should be washed with ______ only. (soap, fresh water)

17. Make sure you know the location of the and the proper use of the ______. (fire hydrant, fire extinguisher)

18. Is it necessary to get first aid for small cuts? (Yes, No)

19. Put out fire in rubbish with ________. (water, carbon tetrachloride)

20. Put out oil fires with ________. (water, CO_2)

21. Electrical fires should be put out with ________. (CO_2, water)

22. It is dangerous to leave tools on the top of a ladder. (Yes, No)

23. Stay clear of the operator's _______ when watching machinery. (vision, safety zone)

24. (True, False) Only use tools for the purpose for which they are designed.

25. When using a screwdriver you should hold the work piece with _____. (your hand, a vise)

26. Always use a hammer with a ______ handle. (tight, loose)

27. You should always hand sharp tools to someone else ______. (blade first, handle first.)

28. (True, False) Securing work frees both hands to operate tools.

29. Should you pull or push a wrench? (Pull, Push)

30. You should leave your tools _____ when finished with a job. (in your tool box, on the work bench)

31. Good practice
    Bad Practice

32. Good practice
    Bad practice

33. Good practice
    Bad practice

34. (True, False) Using files without handles can cause injury.

35. Handles of tools should be kept dry ________. (to prevent rust, to prevent injury)

36. What is one precaution to use when working with metal? (Remove burrs, Ventilate the room)

37. When you see nails sticking out of boards you should ________. (pull them out, ignore them)

38. When carrying long objects, how should you carry the front end? (Low, High)

39. Are gloves a hazard when working around power machinery? (Yes, No)

40. Before using a hammer you should _______. (put on gloves, examine it)

41. (True, False) Always turn power off when finished using a machine.

42. (True, False) It is safe to work in a room where there are gas fumes if the window is open.

43. (True, False) One safe method for testing for a gas leak from a pipe is with a match.

44. The proper way to disconnect an electrical plug is to ______. (pull the wire, grasp the plug)

45. (True, False) When working with electrical switches, you are likely to get a shock if your hands are wet.

46. (True, False) An electric drill should be grounded.

47. It is a safe practice to remove the ground pin from an electric plug when a two-prong receptacle must be used.

48. (True, False) The key can safely be kept in the chuck of an electric drill.

49. Before a drill bit breaks through the work piece you should _____. (stop the drill, decrease pressure.)

50. (True, False) Voltages of less than 200 volts seldom cause serious shocks.

51. (True, False) You can safely work alone on electrical equipment, if the equipment is grounded.

52. What is the safest way to work on electrical equipment when the power is on? (With one hand, With two hands)

53. (True, False) It safe to stand on a wet floor when working on electrical equipment, if you wear safety shoes.

54. A good rule, when working on electrical equipment is ____. (ground your chair, never work alone)

55. (True, False) First, touch a "dead" circuit line with the open hand.

56. Is it possible to get an electrical shock by touching two pieces of equipment at the same time. (Yes, No)

57. (Electrical equipment with frayed cords should ______. (be used and repaired, not be used)

58. Where should you store your soldering iron? (On the bench, In a holder)

59. How should you lift a soldering iron? (By the cord, By the handle)

60. Is it dangerous to reach across a hot soldering iron? (Yes, No)

61. The proper tool to use when repairing printed-circuit boards is a ______. (soldering gun, a low-wattage soldering iron)

62. Remove solder from a soldering iron tip ____ (by flipping the iron toward the floor, by wiping on a soft clean rag)

63. When cutting wires the open-end of the cutter should be pointed ________. (upward, away from anyone)

64. The tip temperatures of a soldering iron or gun can reach _____ degrees. (200, over 500)

65. Which is a safety precaution to take when desoldering a wire? (Wear safety shoes, Check spring tension)

66. How long should you give artificial resuscitation? (10 minutes, Until directed to stop)

67. How soon should a victim be removed from an electrical circuit? (After resuscitation, Immediately)

68. (True, False) Never replace a fuse with one of a higher rating.

69. What may be the results from bridging a fuse? (A fire, Safe operation)

70. What current rating should a replaced fuse or circuit breaker have? (Higher, The same)

71. Safety requires that for using a multimeter one should ________. (wear safety glasses, use the correct range)

72. If you are to remove dust from equipment with an air hose you should ____. (wear safety glasses, wear safety glasses and a dust mask)

73. (True, False) It is safe to touch only one conductor of a power line.

74. Good practice
    Bad practice

75. Good Practice
    Bad practice

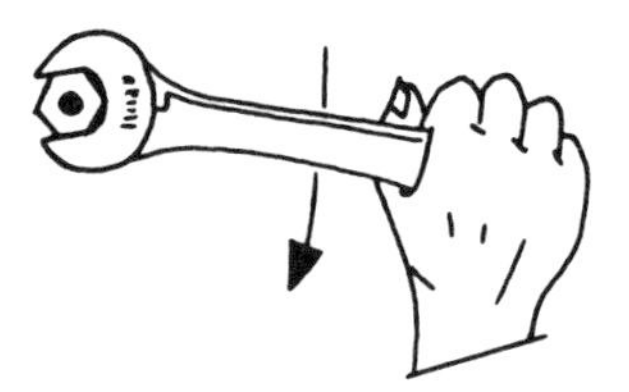

76. Good practice
Bad Practice

77. Good practice
Bad practice

78. Good Practice
Bad Practice

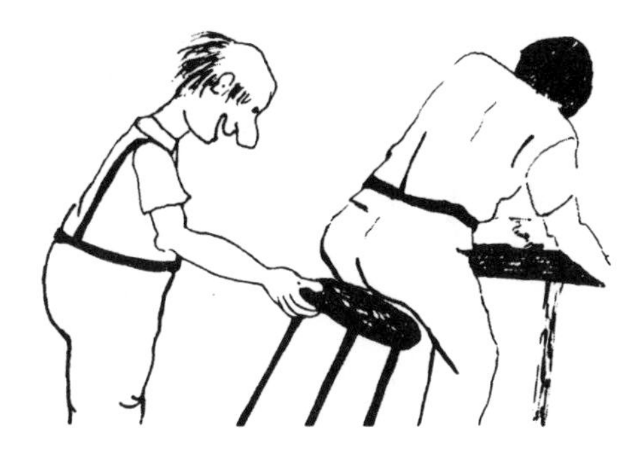

79. Good practice
Bad Practice

81. Good Practice
    Bad practice

Glossary of Terms

Across the Line (Full Voltage)—A method of placing a motor on the full line voltage.

AC—The abbreviation for alternating current. Alternating current that reverses its direction periodically at regular intervals. Value of the voltage is stated in volts and cycles per second (Hertz).

Ambient Temperature—The temperature of the area surrounding a device.

Ampacity—The maximum current rating of a conductor (wire) or device.

Amplifier—A device used to increase a signal voltage or current.

Amplitude—The value of a voltage or current.

Anode—The positive terminal of a device.

Anti-shorting cycling—A control that prevents a compressor from being started within a certain period of time.

Applied Voltage—A voltage applied to a device or circuit.

ASA—American Standard Association

Atom—The smallest part of any element, that comprises the properties of the element.

Attenuator—A device that decreases the value of a signal.

Automatic Ignition—Ignition of an appliance in response to a heat demand.

Automatic Pilot A pilot that operates to ignite the main burners upon heat demand and to shut off the gas supply to the burner if the pilot fails.

Automatic—Self-acting by some control such as temperature or pressure.

Auxiliary Contacts—Small contacts located on motor control relays or contactors that operate other devices.

Base—The part of a transistor located between the emitter and the collector.

Bimetal Strip—A strip comprised of two metals that bends when heated, and is used as one side of a relay.

BTU—Abbreviation for the British Thermal Unit. A commonly used measure of heat. The amount of energy required to raise one pound of water 1 degree F.

Burner—A device that burns an air-gas mixture.

Cable—One or more wires used to conduct electricity.

Capacitance—An electrical component used to shift the phase of a current in a motor.

Capacitor (Run)—A type of capacitor used in the run cycle of a motor.

Capacitor (Start) A type of capacitor used in the starting cycle of a motor.

Capacitor-Start Motor—A motor that uses a capacitor in the starting winding to increase starting torque.

Choke—Another name for an induction coil.

Circuit Breaker—Device to protect a circuit from overload. "Tripped" circuit breaker interrupts circuit when current exceeds a specified amount. See also FUSE.

Circuit Diagram—Drawing using standard symbols to represent path of current from source through switches and components and back to source. Shows the interconnection of wiring and components.

Circuit—Path of electrical current from source supply through wiring to point of use and back to source.

Coil—Another name for an inductor.

Cold Junction—Pertaining to one junction of a thermocouple as opposed to the hot junction.

Comparator—A device or circuit that compares two quantities, such as temperature or voltage.

Component—Electrically operated part such as a switch, coil, relay, motor, thermostat, etc.

Compressor—A device that increases the pressure of a fluid.

Condensing Unit—The component of a refrigeration or air-conditioning unit containing a compressor and condenser.

Conduction—Transfer of heat from one body to another, the transfer of electrical energy, or a current flow.

Conductor—A substance that permits relatively free flow of electric current.

Contactor— A high-current switch device that is controlled by a current.

Contact—Switch, relay, or contactor component that opens and closes to complete or break a circuit.

Continuity—Ability of a circuit to conduct electricity. A complete path.

Controls—Devices used to regulate the movement of a medium such as air, gas, or electricity.

Current Rating—The amount of current that a device is designed to endure.

Current Relay—A relay that is designed to operate on a specific current level. Sometimes used in the starting circuit for compressor motors.

Current—The flow of electrons. The unit of current is the ampere.

DC—Abbreviation for direct current.

Defective Device— A component that does not function properly and which must be replaced.

Delta Connection—A connection of a primary or secondary of a transformer, where the three leads are connected together.

DIAC—A control diode.

Digital Voltmeter—(See DVM)

Diode—A two-element semiconductor rectifier used to convert AC voltage to DC voltage.

DVM—Digital voltmeter. A test instrument for measuring current, voltage, resistance, and for testing diodes.

Dynamic Breaking—A method of using motor current to brake (stop) a motor.

Eddy Currents—Currents induced into the core of a transformer that produce heat and a power loss.

Electric Controller—A device used to control the operation of a motor.

Electrical Interlock—A safety device used to protect equipment or personnel.

Electronic Control—A circuit that uses electronic components to control system action.

Electron—The part of an atom that is current flow.

Emitter—A region of a semiconductor device.

Enclosure—A metal or plastic cover or other surroundings designed to protect equipment or personnel.

Evaporator—A component of a refrigeration system that removes heat from the air or a liquid to transfer heat.

Filter—A device used to remove dust particles from the air, or a device to remove the ripple voltage from a rectifier circuit.

Flue—The passageway through which gasses pass from an appliance to the outer air.

Frequency—the number of cycles of an event in one second or minute. For AC voltage it is the number of cycles (Hertz) per second.

Fuse—Device to protect a circuit from current overload. "Blown" fuse automatically interrupts circuit when current exceeds a specified limit.

Ground—Connection to earth or another connecting body which is connected to earth. Metal components of a system must be grounded for safety to prevent injury.

Heat Anticipator—A component of a thermostat that preheats the sensing element, causing the thermostat to open before the heat selection is reached.

Heat Pump—A heating/air conditioning system that supplies both heat and cooling to an area.

Heat Sink—A metallic device used to increase the area and heat dissipation ability of a semiconductor component.

Hermetic Compressor—A compressor that is air-tight and entirely enclosed.

Holding Current—The value of current necessary to keep a triac or SCR in the ON condition.

Horsepower—The rating units of a motor. 746 watts is equal to one horsepower. In a mechanical system, lifting 1 pound 3300 feet is equivalent to one horsepower.

Hot Junction—The junction of two dissimilar metals in a thermocouple of thermopile.

Hot Wire Relay—A type of relay using a bimetal strip that is used in a motor starting circuit.

Impedance—The total opposition that a circuit offers to a current flow. The unit of impedance is the ohm.

Induced Current—Current induced in a coil, transformer winding, or generator winding by relative motion of the winding and a magnetic field.

Inductor—A coil that opposes a change of current.

Insulator—A material such as glass, rubber, or mica that offers very high resistance to a voltage.

Interlock—A device that prevents an action in an equipment until another event occurs, or a safety device that interrupts the supply voltage when a panel is removed.

Isolation Transformer—A transformer that isolates an equipment or circuit from the primary source and ground.

Jumper—A wire or conductor that connects one point of a circuit to another.

Junction Diode—A diode made up of P and N type semiconductor, with a cathode and an anode.

LED—A light-emitting diode that is usually used as an indicator.

Limit Switch—A switch that confines the movement of a mechanical device to a certain limit.

Load Center—The source of voltage for branch circuits.

Locked Rotor Current—The amount of current that is produced when a motor armature does not rotate.

Lockout Device—A device that prevents the operation of another device.

Low-Voltage Protection—A relay that disconnects a motor from the line voltage when the voltage becomes dangerously low.

Magnetic Contactor—An electromagnetic contactor.

Magnetic Field—The lines of force from a permanent magnet, or a current flow in a wire, coil or transformer.

Maintaining Contact—A set of contacts that holds a relay in the ON position.

Manual Controller—A controller operated by hand.

Micro-Farad—A unit of capacitance. One-millionth of a farad.

Microprocessor—The control section of a computer or a small computer.

Mode—An operating condition.

Motor Controller—An electrical or mechanical device used to start a motor.

Motor—A device that converts electrical energy to mechanical motion.

Multi-Speed Motor—A variable-speed motor.

Negative—An electrical potential, an excess of electrons, or the polarity of one pole of a battery.

NEMA—National Electrical Manufactures Association

NEMA Ratings—The rating of devices by the NEMA.

Neutron—The neutral element of the nucleus of an atom.

Noninductive Load—A load that does not have inductance.

Nonreversing—A device that operates in only one direction. Usually refers to motor action.

Normally Closed (N.C.)—Contacts of a relay or switch that are closed in the OFF position.

Normally Open (N.O.)—Contacts of a relay or switch that are open in the OFF position.

Off-Delay Timer—A timer in which the contact's changes are delayed when the timer is turned off.

Ohmmeter—An instrument used to measure resistance or to check continuity.

Ohm—Measurement unit of resistance (Ω)

On-Delay timer—A timer that delays contact changes when the timer is turned on.

Open (circuit)—Incomplete circuit which cannot conduct electricity.

Operational Amplifier—A high gain semiconductor amplifier.

Oscillator—A device used to develop a square wave or a sine wave voltage.

Oscilloscope—A testing device that displays circuit voltages.

Out-Of-Phase—A voltage or event that has a different time relationship from another.

Overload Relay—A relay that operates to protect a circuit, motor or system in the event of an overload.

Panel board—A panel on which electrical controls are mounted.

Parallel Circuit—A circuit in which all devices or components are connected across each other. A circuit with more than one path for current.

Peak Voltage—The maximum positive or the maximum negative voltage of a waveform.

Peak-To-Peak Voltage—The amplitude of a voltage from one extreme to the other.

Permanent-Split-Capacitor Motor—A single-phase motor in which the start and the running capacitors remain in the circuit after starting.

Phase Shift—A time shift of a voltage or current.

Pilot Device—A device that uses a small quantity of current to control a device that uses a large current.

Pilot Generator—A thermocouple or thermopile used to generate the electrical energy used to operate a pilot safety valve.

Pilot Safety Valve—A gas valve held open by the electrical energy from the pilot generator. Shuts off automatically with the loss of pilot energy.

Pilot—A small gas flame used to ignite the main burner.

Pneumatic Device—A device that operates by a change of air pressure.

Pneumatic Timer—A timing device that operates by a change of air pressure.

Polarity—The positive or negative values of a voltage.

Potentiometer—A variable resistor controlled by a mechanical shaft.

Power Factor—The relationship between the true power in watts to the apparent power in VARS. The cosine Θ of the angle of current to voltage.

Power Rating—The maximum power dissipation rating of a device.

Pressure Switch—A device that changes a switch or switches upon a change of pressure.

Proton—The positive particle in the nucleus of an atom.

Push Button—A switch that is controlled by a pushing action.

Reactance—The opposition that a capacitor or inductor offers to current flow. The unit of reactance is the ohm.

Rectifier—A device that changes AC current into DC current.

Regulator—A mechanical or electrical device that controls an action.

Relay—A mechanical-electrical switching device that controls one or more switching contacts.

Remote Control—The control of an action from a remote position.

Resistance-Start Inductance-Run Motor—A type of split-phase motor that uses the resistance of the starting winding to cause a phase shift that aids the starting of a motor.

Resistance—The opposition that a component or device offers to current flow. The symbol for resistance is R and the unit of resistance is ohm (Ω).

Resistor—A component that limits current or divides voltage.

RMS Value—The effective value of a sine wave voltage.

Run Winding— The winding of a motor that remains in the circuit when the motor is at full speed.

Saturation—The maximum of a quantity that a device can assimilate. The maximum magnetic field of a device or the maximum current of a device.

Schematic—A diagram showing all the parts of a circuit, and usually, the sequence of operation.

SCR—A solid-state rectifier that can be controlled by an external voltage.

Semiconductor—A silicon or germanium device such as a diode or transistor.

Sensing Device—A device that detects a quantity, such as temperature or pressure, and changes it into an electrical signal.

Series Circuit—A circuit in which the components are end-to-end and the same current passes through every component.

Service—The main electrical power source or the main gas source.

Shaded-Pole Motor A motor that uses a small copper or brass ring on the stator poles as a starting device.

Short Circuit—A connection between a hot wire and ground, allowing excessive current to flow with little or no resistance

Short Cycle—When a compressor starts and stops often.

Sine-Wave Voltage The form of an AC voltage developed by a generator.

Slip—The difference between the speed of the rotation of the magnetic field in a motor and the armature.

Snap Action—The action of an overload device to quickly open a circuit.

Solenoid Valve—A valve that is controlled by solenoid action.

Solenoid—A device comprised of a coil and a movable iron core. The basics of a contactor.

Solid State—Usually refers to semiconductor devices such as diodes, and transistors, etc. or other devices that have no moving parts.

Split-Phase Motor—A motor that uses capacitors or coils to cause an effective two-phase system.

Staging Thermostat—A thermostat that senses temperature change and the need for more or less heating or cooling.

Starter—A device to aid in starting a motor.

Starting Winding—A winding in a motor that is engaged during the starting cycle to aid in overcoming inertia.

Stator—The stationary part of a motor or relay. The case of a motor.

Step-Down Transformer—A transformer with less winding on the secondary than the primary, that produces an output voltage less than the input voltage.

Step-Up Transformer—A transformer that steps up the primary voltage at the secondary. The power of the primary is equal to the power of the secondary. Therefore, the secondary current is stepped down in the direct ratio of the rise in secondary voltage.

Surge—A sudden increase in pressure or voltage.

Switch—Device to turn on and off an electrical circuit.

Synchronous Speed—A constant speed.

Temperature Relay—A temperature-controlled relay.

Terminal—Connection point between wiring and electrical components. Also, ring or push-on connector.

Test Probes—Metal components of VOM test leads.

Thermistor—A semiconductor resistor that changes value with temperature. An increase in temperature decreases its resistance.

Thermocouple—A thermo-electric device that converts heat energy into electrical energy.

Thermostat—Device for controlling temperature levels in air-conditioning, refrigeration, and heating systems

Three-phase Voltage—Three voltages that are 120 degrees out of phase.

Torque—The turning power of a motor.

Transducer—A control device.

Transformer—Device for raising or lowering AC voltage.

Transistor—A solid-state device used in amplifiers or control circuits.

Triac—A bidirectional rectifier used to control current.

Troubleshooting—The act of evaluating a defective circuit or device.

Valence Electrons—The electrons in the outer shell of an atom.

Variable Resistor—A resistor whose value can be changed from a low to high limit.

Voltage Drop—The voltage that is developed across a component or device due to resistance and current.

Voltage Rating—The correct voltage that should be applied to a device for correct operation.

Voltage Regulator—A device that controls the level of a voltage.

Voltage—The difference of potential between two points.

Voltmeter—A device to measure voltage.

Volt—Measurement of electrical pressure.

VOM—A Volt-Ohm-Milliammeter used to measure voltage current, and resistance.

Watt—The unit of power and product of current and voltage.

Waveform—The shape of a voltage as it changes. An AC voltage is a sine wave.

Wye Connection—The connection of the windings of a three-phase transformer or a three-phase load.

Zener Diode—A diode that is used as a voltage regulator.

Bibliography

Modern Refrigeration and Air Conditioning, Althouse, Turnquist, and Bracciano, GoodHeart and Wilcox Company, Inc., 1992

Electrical Theory and Control Systems in Heating and Air Conditioning, Dorner, and Greenwald,. Delmar Publishing Inc., 1994

HAVC Control Systems, Schneider, Raymond, John Wiley Publishers, 1981

Electricity for Air Conditioning and Refrigeration Technicians, Mahoney, Edward, Prentice Hall, Inc.

Electricity and Controls for Heating, Ventilation, and Air Conditioning, Herman and Sparkman, Delmar Publishers, Inc., 1991

Servicing Environmental Systems, Sheet Metal and Air Conditioning Industry Union, 1995

Periodicals

Western HVACR News, P.O. Box 42749, Los Angeles, CA 90050-0749

Honeywell Climate News, Honeywell, Inc.

RSEJ Journal, 9560 SW Nimbus Avenue, Beaverton, OR 97008

Refrigeration Service and Contracting, P.O. Box 7021, Troy, MI 48007-9916

Index